QUANTUM ELECTRONICS IN LASERS AND MASERS

KVANTOVAYA RADIOFIZIKA

КВАНТОВАЯ РАДИОФИЗИКА

The Lebedev Physics Institute Series

Editor: Academician D. V. Skobel'tsyn

Director, P. N. Lebedev Physics Institute, Academy of Sciences of the USSR

Proceedings (Trudy) of the P. N. Lebedev Physics Institute

Volume 31

QUANTUM ELECTRONICS IN LASERS AND MASERS

Edited by
Academician D. V. Skobel'tsyn
Director, P. N. Lebedev Physics Institute
Academy of Sciences of the USSR, Moscow

Translated from Russian

SPRINGER SCIENCE+BUSINESS MEDIA, LLC

1968

The original Russian text, published by Nauka Press in Moscow in 1965
for the Academy of Sciences of the USSR as Volume XXXI of the Proceedings (Trudy) of the P. N. Lebedev Physics Institute, has been corrected
by the editor for this edition.

Квантовая радиофизика

Труды Физического института им. П. Н. Лебедева
Том XXXI

Library of Congress Catalog Number 68-13059

ISBN 978-1-4684-9974-2   ISBN 978-1-4684-9972-8 (eBook)
DOI 10.1007/978-1-4684-9972-8

CONTENTS

CONTENTS

METHODS OF OBTAINING NEGATIVE TEMPERATURE STATES IN SEMICONDUCTORS

Yu. M. Popov

INTRODUCTION

In the beginning of the fifties, based on the utilization of induced transitions in quantum systems, a new theory concerning the generation of coherent electromagnetic waves was developed in the USSR and the USA. In the mid-fifties, the first molecular generators of coherent radiation in the microwave range (masers) based on this theory were created. This period is now generally regarded as the genesis of a new branch of physics, quantum radiophysics. During the last ten years, the development of quantum radiophysics has resulted in major scientific and technical achievements. Paramagnetic quantum amplifiers operating in the centimeter and decimeter wavelength range, as well as ultrastable frequency generators using molecular and atomic beams were designed in those years and are now enjoying widespread application.

A distinct new stage in the development of quantum radiophysics was heralded by the creation in 1960 of quantum oscillators operating in the optical and infrared ranges (optical and infrared masers, or lasers), which use luminescent crystals and gas mixtures as the working material. Research on the generation of coherent waves at optical wavelengths blossomed so rapidly that generation has been realized in more than twenty forms of luminescent crystals and more than ten different gases and gas mixtures during the last three years, while the number of different spectral lines used for generation has waxed to more than a hundred.

Very recently, the generation of coherent light has been accomplished through the medium of semiconductors, although the theoretical groundwork for its possible realization was advanced four years previously. Semiconductor quantum oscillators occupy a special place among generators of optical and infrared radiation.

Semiconductor quantum oscillators can be used directly for the transformation of electrical energy into coherent radiation, the efficiency of such transformation approaching 100% under certain conditions (see §2, Chapter II).

Semiconductors are capable of yielding, with regard to transmitted radiation, a gain equal in absolute value to the absorption coefficient, the magnitude of which is generally many times larger in semiconductors than the corresponding gains in other substances used for lasers. Semiconductor quantum oscillations may therefore be imparted very small dimensions (fractions of a millimeter).

---

*Dissertation presented toward fulfillment of the requirements for the academic degree of Doctor of Physicomathematical Sciences, 1963.

In the radiative recombination of current carriers in a semiconductor, the position of the energy level in a band can change, depending on the magnitude of the magnetic field, which means that the frequency of the generated light can be varied by means of a sufficiently strong magnetic field. Since the effective mass of the carrier is small in a number of semiconductors, the variation of frequency by a magnetic field in a semiconductor is potentially more effective than in crystals utilizing transitions between impurity levels.

Already there are a variety of semiconductors (germanium, silicon, etc.) and semiconductor compounds (compounds of the groups III and V, II and VI) that offer promise for the creation of coherent light sources yielding radiation from the far infrared to the ultraviolet. The use of semiconductors to build lasers encompassing a broad range of frequencies depends largely on the ability to prepare various pure semiconductor materials with forbidden bands of different widths and on the ability to use various impurities in semiconductors. The effective operation of such oscillators requires the solution of a whole series of technological problems: the preparation of optically homogeneous materials, the fabrication of multilayer p−n junctions and structures of the type p−i−n, high precision in the finishing of surfaces and the deposition of special coatings to enhance or diminish the reflection coefficient.

Among the drawbacks of semiconductors as a material for lasers by comparison with other materials, especially gases, it is essential to cite the large energy width of the recombination radiation spectrum, the order of magnitude of which in semiconductors is generally equal to that of the thermal energy of the carriers, whereas in the case of gases this width is due to the Doppler effect and comprises but an insignificant part of the energy of thermal motion. For this reason, the induced emission state is attained in semiconductors only at relatively high excitation densities, and there must not be appreciable heating of the crystal lattice. The simultaneous fulfillment of these requirements is the most difficult problem in the fabrication of lasers using semiconductors.

As is generally known, it is necessary for the creation of quantum oscillators to transfer the working substance into such a state that it is capable of amplifying electromagnetic radiation (negative temperature state). The negative temperature state may be obtained in semiconductors by means of various excitation sources: an electric current, a beam of ionized particles (fast electrons), optical illumination from powerful light sources. Under definite excitation conditions, a semiconductor laser can function in a continuous mode, and the radiation can be modulated by modulation of the current passing through the sample.

The proposed use of semiconductors for quantum oscillators at optical wavelengths (lasers) was first advanced in 1958 [1] and was one of the first proposals for using induced transitions to create sources of coherent radiation in the optical range. Justification was given in that investigation for the possibility of obtaining in semiconductors thermodynamically nonequilibrium states capable of amplifying electromagnetic waves. For the realization of such states, it was proposed that a homogeneous and sufficiently pure semiconductor be driven by electric field pulses. Upon rapid removal of the field in this method of excitation, it is possible to obtain the negative temperature state essential for generation, provided the temperature of the crystal is adequately low.

In 1960, in [2], the notion was advanced of using indirect transitions in semiconductors for the purpose of generating negative temperature states. In later publications, the same authors [3, 4] analyzed the conditions for obtaining a negative absorption coefficient, which is particularly important in the case of indirect transitions with their lower probability by comparison with direct transitions.

Early in 1961, a method of injecting minority current carriers through a p−n junction in degenerate semiconductors was proposed in [5] for the purpose of obtaining negative temperature states in semiconductors in a narrow layer near the p—n junction. This was subsequently the method of excitation used in building the first semiconductor oscillators.

At the First and Second International Conferences on Quantum Electronics [6, 7], it was suggested that transitions between the energy levels occurring in strong magnetic fields (Landau levels) and the transitions between levels of certain impurities in semiconductors be used for generation in the submillimeter and far-infrared regions. However, the very small relaxation times at these levels and the presence of absorption in transitions to higher levels, create difficulties in the realization of generation using such levels. Moreover, as shown in [8], due to the reduction in relaxation time with the lattice for current carriers existing at the higher energy levels, it is practically impossible to obtain a negative temperature state between levels belonging to a common zone.

The proposal to use the negative effective mass of carriers for the amplification of electromagnetic radiation [9] was subjected to criticism in [10-12], where it was shown that a necessary condition for amplification is the formation of states with negative temperature.

Experimental studies have been in progress since 1959 at the P. N. Lebedev Institute of Physics of the Academy of Sciences of the USSR (FIAN SSSR) on the creation of negative temperature states in semiconductors for the purpose of designing lasers. The method of excitation of a homogeneous semiconductor by a strong electric field was employed in [13, 14], wherein the behavior of a nonequilibrium electron—hole plasma was investigated in pure InSb. In this paper, apparently, only indirect demonstrations were obtained of the formation of negative temperature states (high carrier concentration) at certain instants of time, but the authors failed to obtain direct evidence of the presence of an inverted population with respect to narrowing of the emission line or of the directivity of the radiation.

In [15], a study was made of the negative photoconductivity in certain silicon samples with the hope of detecting negative temperature states in indirect transitions. It is likely that strong absorption by free carriers hampered the realization of a negative absorption coefficient.

Recombination radiation at large current densities in the forward direction through an InSb tunnel diode was investigated in [16] for the purpose of obtaining negative temperature states by the method of injection through a p—n junction. As shown by estimates presented by the authors of that article, a negative temperature state was indeed obtained at large current densities. However, they performed no spectral measurements to provide direct proof of the presence of negative absorption.

Radiative recombination corresponding to injection through a p—n junction in GaAs was observed in [17]. The authors investigated the presence of an inverted population as one possible explanation for the narrowing of the emission line noted in their work.

Toward the end of 1962 and the beginning of 1963, a number of firms in the USA [18, 19] and FIAN SSSR [20] succeeded in obtaining a strong narrowing of the emission line and realizing generation in a p—n junction in GaAs at current densities of about 10^4 A · cm^{-2} and a temperature of 78°K. Last year (1964), the semiconductor compounds InAs, InP, and solid solutions of GaAs and GaP, InAs and GaAs were used to make quantum oscillators yielding radiation from the red part of the visible spectrum up to wavelengths of 3.2 μ.

In the present article, we investigate various methods for obtaining states with negative temperature in semiconductors. By analysis of the interaction of electromagnetic radiation with matter, the conditions are derived for formation of the thermodynamically nonequilibrium states in semiconductors (negative temperature states) essential to the amplification of electromagnetic radiation. As there are a number of different methods for the excitation of a semiconductor (electric current, ionizing beams, optical pumping), we investigate in each specific case of excitation the possible routes by which the necessary conditions are fulfilled. Special attention is devoted to the energy efficiency of a particular mode of excitation.

The hypothetical possibility of creating negative temperature states in semiconductors depends on the relation between the slowing-down time and lifetime of the nonequilibrium current carriers. We will therefore solve the kinetic equation, taking into account Fermi degeneracy, in order to obtain the slowing-down times, which may be shorter than the lifetime in most semiconductors.

Considerable emphasis is laid on analyzing the possible utilization of indirect transitions in semiconductors for the amplification of radiation. As a rule, the probability of indirect transitions is much lower than the probability of direct transitions. For the case of indirect transitions, therefore, we take absorption processes into account and derive the conditions for obtaining a negative absorption coefficient.

The first chapter of the article is devoted to the method of obtaining negative temperature states with the excitation of a homogeneous semiconductor by a pulsed electric field. In this chapter the condition is derived for obtaining thermodynamically nonequilibrium states capable of amplifying electromagnetic radiation; consistent with conventional terminology, such states are called negative temperature states. These conditions may be written in the case of semiconductors as inequalities on the distribution functions or as conditions on the energy separation of the momentarily noncoincident Fermi quasi-levels for electrons and holes. The possibility of obtaining negative temperature states in interband transitions and in transitions to an impurity is analyzed. At the end of the chapter, we discuss the difficulties encountered in obtaining negative temperature states with such a mode of excitation and indicate the advantage of using possibly shorter-duration electrical pulses to increase the efficiency.

In the second chapter, we consider the creation of negative temperature states by the injection of minority current carriers into the region near a p−n junction in a degenerate semiconductor. In this case, fairly small voltage biases are required in the forward direction in order to obtain an inverted population in the region of the p−n junction. The narrowing of the emission line is analyzed and the condition for self-excitation derived in the presence of a negative temperature state. The theoretical calculations presented in this chapter suggest the conditions under which the efficiency may be made to approach 100% by the injection of minority carriers through a p−n junction.

The third chapter is given over to the problem of using indirect transitions in certain semiconductors (for example, germanium and silicon) for the amplification of electromagnetic radiation. Upon indirect transition, a phonon is emitted simultaneously with a photon, i.e., the system is analogous in a certain sense to a three-level system in which an optical transition occurs at a level separated from the ground state by a distance equal to the energy of the emitted phonon. Clearly, at a sufficiently low crystal temperature, this level becomes unoccupied, and the relatively low nonequilibrium population of the upper level leads to the amplification of transmitted electromagnetic radiation, provided other absorption mechanisms are removed. Such systems are similar to systems utilizing the Raman effect for the amplification of electromagnetic radiation, when the medium is capable of amplifying radiation while remaining in the state of thermodynamic equilibrium. However, the probability of indirect transitions is small in comparison with direct transitions (without the participation of phonons). The gain obtained by means of indirect transitions is small, therefore, and it is necessary to take into account absorption processes before a final decision can be made as to the possibility of obtaining a negative absorption coefficient. The theoretical calculations show that the inevitable absorption by free carriers − the main obstacle to the procurement of a negative absorption coefficient − also sets a demand for large excitation densities for the amplification of radiation. The excitation intensity may be reduced considerably if the indirect transition originates from the exciton state, because then the transition probability is greatly increased, hence the gain is also enhanced.

In the fourth chapter we look into the possibility of obtaining negative temperature states in semiconductors excited by an electron beam with energies of tens and hundreds of kiloelectron volts or by powerful optical pumping, including pumping by radiation from optical quantum oscillators and oscillators with a modulated quality factor. Although the method of current carrier injection through p−n junctions in degenerate semiconductors is capable of yielding efficiencies close to 100%, the interval of generation for this mode of excitation is still quite narrow. The use of electron beams should increase the thickness of the excited layer to fractions of millimeters and thus enhance the over-all radiation power. But, as inferred from the theoretical calculations of this chapter, the maximum energy yield of a semiconductor quantum oscillator with excitation by an electron beam cannot be higher than 40%.

In the fifth chapter we calculate the slowing-down times of nonequilibrium current carriers, which are not in equilibrium thermodynamically with the lattice inside the corresponding bands. Since in nearly all the excitation methods considered the electrons and holes are initially found inside a wide energy interval in the corresponding bands, a negative temperature state is only possible in the event that the electrons and holes are able, within their lifetime, to occupy a narrow energy belt within their zones, i.e., if the slowing-down time is shorter than the lifetime. For the calculation of the slowing-down times, we derive a kinetic equation describing the interaction of mobile current carriers with the acoustic and optic modes of the semiconductor lattice, taking into account the possible degeneracy of the carriers. It follows from the calculations of this chapter that the slowing-down of nonequilibrium current carriers to energies on the order of the optical phonon energy occurs rather quickly in interaction with optical phonons; further retardation occurs upon interaction with acoustic phonons and depends on the mobility of the carriers upon scattering by the acoustic vibrations of the lattice.

The sixth chapter gives the form of the distribution function for nonequilibrium current carriers inside the corresponding zone in the case of a monochromatic source, for example, radiation from a laser. The difficulties incurred in the generation of an inverted population are pointed out for such a mode of excitation between any two levels indigenous to a common band.

We conclude with a comparison of the various methods for realizing negative temperature states in semiconductors and a discussion of certain as yet unresolved problems associated with the application of semiconductors for the generation of coherent light. Although the article is aimed at a theoretical analysis of the generation of negative temperature states in semiconductors, the final results are compared with the experimental data. The major portion of the theoretical investigations was completed prior to the beginning of the current extensive program of experimental research on the generation of coherent light by means of semiconductors, at a time when the potential realization of this phenomenon was still highly controversial. The creation of the first semiconductor oscillators demonstrated the validity of many of the theoretical notions developed in this article and gave impetus to new theoretical and experimental problems of a concrete nature, some of which are mentioned.

THE GENERATION OF NEGATIVE TEMPERATURES BY THE PULSED ELECTRIC FIELD EXCITATION OF HOMOGENEOUS SEMICONDUCTORS

1. Negative Temperature

The concept of negative temperature is now widely used in describing the operation of quantum oscillators and amplifiers, because systems found in a negative temperature state amplify electromagnetic radiation transmitted through them by virtue of induced emission.

The conditions under which the onset of "negative absorption" was possible for radiation were first formulated in [21].

The concept of negative temperature was first introduced in [22] as a device characterizing the nonequilibrium distribution of Li nuclei with respect to the Zeeman levels in the case when the number of nuclei at the higher level exceeded the number of nuclei at the lower level.

We investigate a certain set of identical particles — atoms, molecules, electrons — in a semiconductor.

In the state of thermodynamic equilibrium, the particle distribution function f_i with respect to the energy levels ε_i is defined by the equations of statistical physics:

for particles with integer spin, by the Bose−Einstein formula:

$$f_i = (e^{\frac{\varepsilon_i - \mu}{kT}} - 1)^{-1}, \tag{1}$$

for particles with half-integer spin, by the Fermi−Dirac formula:

$$f_i = (e^{\frac{\varepsilon_i - \mu}{kT}} + 1)^{-1}, \tag{2}$$

where μ is the chemical potential, k is the Boltzmann constant, and T is the temperature. In the case $f_i \ll 1$, Eqs. (1) and (2) go over to the classical Boltzmann distribution:

$$f_i = e^{\frac{\mu - \varepsilon_i}{kT}}. \tag{3}$$

It follows from Eqs. (1)-(3) that the probability of filling an energy level diminishes, the higher the level, i.e., $f_i < f_j$ if $\varepsilon_i > \varepsilon_j$.

In the case of external influences acting on the system, the form of the distribution function may change considerably from that of Eqs. (1)-(3). If now we formally describe the ratio of the distribution functions for two different levels, ε_i and ε_j, by the following expression:

$$\frac{f_i}{f_j} = e^{-\frac{\varepsilon_i - \varepsilon_j}{kT_{ij}}}, \tag{4}$$

where $\varepsilon_i > \varepsilon_j$, each pair of levels being characterized, of course, by its own effective temperature T_{ij}, then all systems described by the distribution functions may be divided into two classes: (1) systems for which the probability of particles existing at a level with energy ε diminishes with increasing energy and for which the inequality $f_i < f_j$ holds if $\varepsilon_i > \varepsilon_j$ over the entire interval of possible energy states; and (2) systems for which the inequality $f_i > f_j$ holds if $\varepsilon_i > \varepsilon_j$ for at least two energy levels.

It is evident that the effective temperature, determined according to (4) for systems of the first class, will be positive for any pair of levels, while for systems of the second class, at any rate for a pair of levels ε_i and ε_j, the effective temperature must be negative. Consequently, the states of systems belonging to the second class have come to be called negative temperature states.

In the interaction of electromagnetic radiation with a substance, the atoms of the latter may, on the one hand, absorb radiation quanta and transfer to a higher energy state. On the other hand, atoms found in an excited energy state may emit radiation quanta and transfer to a lower energy state. The emission of a quantum of energy by an excited atom occurs in two ways: by the spontaneous transition of an atom to a lower level due to interaction with the zeroth vibrational modes of the electromagnetic field, or by induced transition to a low energy level due to the external electromagnetic field acting on the atom. In this latter case, the atom emits a quantum that is completely identical to the quanta producing induced transition. Hence, upon interaction of the excited atoms with an electromagnetic field of a certain polarization frequency and direction of propagation, the number of field quanta will be increased by the induced emission. Spontaneous transitions of excited atoms takes place irrespective of any external radiation. As a result, spontaneous transitions have no bearing on the process of electromagnetic wave amplification and determine the noise in quantum systems.

Induced emission was first postulated by Einstein in 1917 in his investigation of the problem of thermodynamic equilibrium of emission from matter [23].

It will be apparent from the ensuing discussion that systems of the first and second classes interact with electromagnetic radiation differently, which is what motivated this segregation into classes in the first place.

Let us consider the interaction of particles subject to Bose or Fermi statistics and having two energy levels ε_i and ε_j ($\varepsilon_i > \varepsilon_j$) with radiation at a frequency $\omega_{ij} = (\varepsilon_i - \varepsilon_j)/\hbar$ (where $\hbar$ is Planck's constant).

The number of photons emitted by the system in unit time during transition from the state i to the state j is equal to

$$I^+ = w_{ij}(n_\omega + 1)f_i(1 \mp f_j), \tag{5}$$

where w_{ij} is the probability of transition from the state i to the state j with the emission of a photon and n_ω is the number of photons in the radiation oscillator ω. The minus sign in the second term corresponds to Fermi statistics, the lower (plus) sign to Bose statistics.

The number of photons absorbed by the system per unit time is equal to

$$I^- = w_{ji}n_\omega f_j(1 \mp f_i), \tag{6}$$

where w_{ji} is the probability of an atom going from the state j to the state i with the absorption of a photon.

According to the principle of detailed balancing [24], the probabilities of the direct and reverse processes are equal, i.e., $w_{ij} = w_{ji}$. Subtracting (6) from (5), we obtain the excess of the number of transitions with emission of quanta over the number of transitions associated with absorption:

$$I^+ - I^- = w_{ij}[n_\omega(f_i - f_j) + f_i(1 \mp f_j)]. \tag{7}$$

The first term of Eq. (7), which is proportional to the number of photons, is related to induced emission or resonance absorption, whereas the second term in (7) does not depend on the number of photons and gives the spontaneous transitions in the system. In order for the system to amplify electromagnetic radiation transmitted through it at a frequency ω, it is necessary that the following condition be satisfied, according to (7):

$$f_i - f_j > 0, \tag{8}$$

i.e., the system must be in a state with negative temperature relative to the levels ε_i and ε_j. The condition (8) is a necessary but not in general a sufficient condition for amplification, because transitions are possible in the system with a given frequency ω between other pairs of levels, relative to which the temperature is positive.

Consequently, the sufficient condition for amplification is

$$\sum_{i,\,j} w_{ij}(f_i - f_j) > 0, \tag{9}$$

where the summation is taken over all pairs of subscripts for $\varepsilon_i - \varepsilon_j = \hbar\omega$.

Equation (9) essentially represents the condition for the occurrence of a negative absorption coefficient and is stronger than the condition for the existence of negative temperature. Clearly, if a negative temperature occurs for any pair of levels ε_i and ε_j, then (9) is fulfilled automatically, but even in the case when the temperature has a positive value for some of the energy levels, the inequality (9) can still be fulfilled. The latter case is the one observed in semiconductors, whereas in gases or luminescent crystals the case is often realized in which only the separation between two certain levels satisfies the condition (9), while all other levels drop out of the sum due to their failure to meet the condition $\hbar\omega = \varepsilon_i - \varepsilon_j$. For the conditions of amplification of electromagnetic radiation, this result applies to the simplest process, when only one type of particle, in this case a photon, is emitted upon a quantum transition taking place in the system.

In more complicated situations when several types of particles take part in the processes simultaneously, it is possible to obtain states capable of amplifying electromagnetic radiation by external influences acting on the additional "degree of freedom" of the system that occurs in this case.

The conditions for amplification for one such process, indirect recombination in a semiconductor, is discussed in §1 of Chapter III.

2. The Generation of Negative Temperatures in Homogeneous Semiconductors Excited by Electric Field Pulses

The realization of negative temperatures in interband transitions by the method of exciting a homogeneous semiconductor with electric field pulses was proposed in [1].

For a sufficiently high voltage applied to a homogeneous or impurity-free semiconductor, a sharp increase in the number of mobile carriers takes place, of electrons in the conduction band and holes in the valence band in the material. Two factors cause the increase in nonequilibrium carriers: (1) impact ionization of the valence band by a carrier if in the electric field it has accumulated enough energy to create an electron−hole pair; and (2) the Zener (or tunnel) effect, which may be of considerable importance in the case of narrow forbidden bands or when using very thin semiconducting films.

The mechanism of impact ionization in semiconductors has been investigated in detail in [25-30], the theory of the Zener effect in [31]. We will not consider the mechanism and causes of the increased nonequilibrium current carriers in detail. We merely point out that, according

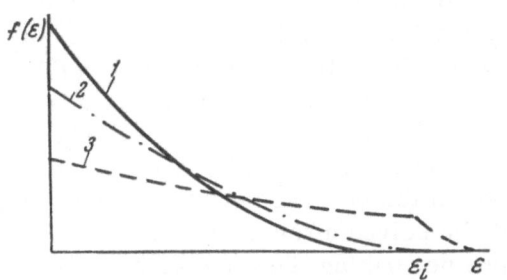

Fig. 1. Dependence of the distribution of conduction electrons with respect to energies on the strength of the external electric field. The curve numbers increase with the field strength. After the ionization threshold ε_i, the distribution function decays exponentially.

to the references cited above, the breakdown of p-type semiconductors occurs in fields such that the mean energy of electrons in the conduction band or holes in the valence band are equal in order of magnitude to the ionization potential. The magnitude of the field F_i for which the mean electron energy is of the same order as the ionization energy, $\varepsilon \approx \varepsilon_i$, is determined from the condition

$$F_i \approx \frac{(m\hbar\omega_0)^{1/2}}{e\tau_{op}(\varepsilon_i)},\qquad(10)$$

where m is the effective mass of the carrier, ω_0 is the frequency of the optic vibrations of the lattice, and τ_{op} is the relaxation time of the carrier in interaction with the optic modes of the lattice.

As shown in §1, a negative temperature state between the levels ε_i and ε_j ($\varepsilon_i > \varepsilon_j$) is attained under the condition†

$$f(\varepsilon_i) > f(\varepsilon_j).\qquad(11)$$

In the investigated interband transitions under the condition (11), the energy ε_i is ascribed to some level of the conduction band, ε_j to a level of the valence band. In the above-cited references [25-30], a distribution function is also given for electrons of the conduction band (holes of the valence band) in a strong electric field during scattering by vibrations of the lattice. For example, according to [28], it has the form (Fig. 1):

$$f(\varepsilon) = N'' \exp\left[-\frac{\varepsilon^2}{2p'(kT)^2}\right],$$

where

$$p' = \frac{(eFla)^2}{6mu^2kT}(1+R)^{-1} + R\frac{(\hbar\omega_0)^2}{mu^2kT},\qquad R \approx \frac{D^2}{C^2},\qquad l_a = \frac{9\pi M u^2 \hbar^4 n_0}{4C^2 m^2 kT}.\qquad(12)$$

Here F is the strength of the external electric field, u is the velocity of sound in the semiconductor, M is the mass of an atom of the lattice, n_0 is the density of lattice atoms, k is the Boltzmann constant, T is the lattice temperature, and C^2 and D^2 are the constants of interaction with the acoustic and optical modes of the lattice, respectively.

We note in Eq. (12) that the energy ε is read upwards from the bottom of the conduction band in the case of electrons, downwards from the top of the valence band for holes. It is apparent from Eq. (12), as well as visibly from Fig. 1, that the distribution function for electrons $f_e(\varepsilon)$ reaches its largest value for $\varepsilon = 0$; an analogous result applies to the hole distribution function $f_h(\varepsilon) = 1 - f_e(\varepsilon)$, which reaches its maximum at the top of the valence band.

Consequently, the most favorable conditions for the realization of negative temperature states in interband transitions occurs between levels intrinsic to the edges of the corresponding bands, i.e., if we regard ε_i in (11) as the bottom level of the conduction band, ε_j as the top

---

† Here we are considering only direct transitions, i.e., the case when only one particle, a photon, is emitted on recombination of an electron and hole. Indirect transitions will be taken up in Chapter III.

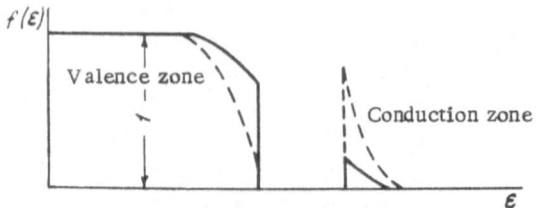

Fig. 2. Energy distribution of electrons in a
semiconductor. The solid curve represents
the thermodynamic equilibrium distribution.
The dashed curve represents the thermodynamic nonequilibrium distribution corresponding to the presence of negative temperature
between the valence and conduction bands.

level of the valence band. Without an external
electric field, the distribution function is determined by the well-known expression (2),
which does not meet the condition (11). The
problem of the possible fulfillment of (11) for
levels ε_i and ε_j not belonging to the edges of
the bands in the presence of an electric field,
can only be solved after the system of kinetic
equations describing the processes occurring
in a semiconductor has been analyzed. It appears to us that an arbitrarily high value of the
electric field is equivalent to a high temperature, i.e., it leads to a function $f(\varepsilon)$ that decreases monotonically with the energy, tending
in the limit to a constant, a situation which corresponds to systems of the first class con-

sidered in §1. We assume, therefore, that it is impossible with such a mode of excitation to obtain negative temperature states without removal of the external field or without some other
mechanism enabling the field-induced carriers to compress within a narrow energy strip near
the bottom of the conduction band and top of the valence band, or to increase their concentration appreciably (for example, by means of the "pinch effect" — see below). The fact is that,
even though the number of mobile electrons and holes has increased in the corresponding bands
in a strong field, they still acquire energy from the external field and are distributed over a
wide energy belt of the corresponding band (they occupy the energy interval from the bottom of
the band to the energy characterizing the ionization threshold).

After sufficiently rapid removal of the external electric field at a low temperature of the
crystal lattice, the nonequilibrium carriers undergo the following two processes: (1) slowing
down inside the corresponding bands to an energy kT (where T is the temperature of the crystal
lattice) with a time t_s; and (2) recombination of carriers of opposite sign with a time τ_c.

Only in the case $t_s \ll \tau_c$ is the formation of negative temperature states possible. In fact,
when this condition is fulfilled, the electrons and holes occupying a wide energy belt in an external field become compressed into a narrow belt of width on the order of kT near the bottom
of the corresponding bands upon removal of the field during a time shorter than τ_c, as is graphi-
cally illustrated in Fig. 2.

Electrons and holes arriving in thermodynamic equilibrium with the lattice may be characterized by the Fermi quasi-fields, which have different values by virtue of the nonequilibrium
value of the concentrations. The Fermi quasi-level μ_e for electrons is defined as follows:

$$\int_0^\infty \left(e^{\frac{\varepsilon-\mu_e}{kT}} + 1\right)^{-1} \rho_e(\varepsilon)\, d\varepsilon = n_e, \tag{13}$$

where $\rho_e(\varepsilon)$ is the density of the levels in the conduction band, and the integral in (13) is taken
over all energy levels of the conduction band; n_e is the nonequilibrium concentration of electrons in the conduction band.

The definition of the Fermi quasi-levels for holes is analogous, except that the energy is
read in the opposite direction. We assume now that the distribution for slowed electrons and
holes is a Fermi distribution with lattice temperature T, but with the corresponding Fermi
quasi-levels μ_e and μ_h.

Considering in this case the most favorable conditions for negative temperatures between the bottom of the conduction band and top of the valence band, we arrive at the condition

$$\mu_e + \mu_h \geqslant \Delta, \tag{14}$$

where Δ is the width of the forbidden band.

Consequently, in order to obtain negative temperature states in direct transitions between bands, it is necessary that the energy separation of the Fermi quasi-levels be greater than the width of the forbidden band.

The necessary concentration of nonequilibrium carriers and temperature of the semiconductor lattice for the formation of negative temperature states are determined from the conditions (13) and (14), provided only that $t_S \ll \tau_C$. In the fifth chapter a derivation of the slowing-down time is given with regard for interaction with acoustic and optical phonons, and it is shown that the condition $t_S < \tau_C$ is fulfilled for almost all known semiconductors, although the lifetime τ_C decreases rapidly with such large nonequilibrium concentrations, clearly as the result of triple collision processes, which are analyzed in detail in a special appendix. It is important to point out that the lifetimes of the nonequilibrium carriers in the case of interband recombination in pure crystals are generally many times larger than the lifetimes of nonequilibrium carriers in impurity semiconductors.

In the case of ionization of the valence band, the magnitude of the electric field required in order to form a high enough concentration rapidly increases with the width of the forbidden band. For semiconductors with large forbidden bands, therefore, high field strengths in relatively short pulses are needed. If the effective masses of the carriers in the semiconductor are not small in comparison with the electron mass, the nonequilibrium concentrations required for degeneracy [for fulfillment of the condition (14)] are high. In this case, it is difficult to ensure a low lattice temperature because of the large Joule heat losses.

The creation of negative temperatures is facilitated in the given method when semiconductors with a narrow forbidden band, large mobility, and low effective mass of the carriers is used, if the lifetime of the nonequilibrium carriers in them is sufficiently high. For this reason, in [13, 14] a homogeneous InSb semiconductor was excited with short-duration pulses, giving a high density of electrons and holes produced as the result of impact ionization. The temperature of the sample was 78°K and, in view of the brevity of the pulses, the sample did not become overheated, even at high currents. For a field strength greater than 200 W · cm^{-1}, the recombination radiation spectrum of nonequilibrium electrons and holes created by the field was observed in the vicinity of 4-6 μ.

The recombination radiation spectrum indicated that the total temperature of the electrons and holes was near 500°K. Observation of the dimensions of the luminescence region yielded direct proof of the existence of magnetic constriction of a luminescent filament ("pinch effect") with increasing current through the sample, so that the diameter of the luminescent filament at a current of 50 A was about 0.2 mm, thus corresponding to a current density of 10^5 A · cm^{-2}. At this current density and a drift velocity equal to $3 \cdot 10^7$ cm · sec^{-1}, the average electron (hole) concentration in the filament is $5 \cdot 10^{16}$ cm^{-3}. Magnetic constriction of the filament can in principle raise the nonequilibrium carrier concentration in the filament to the extent that the condition (14) could be satisfied even in the presence of a field. It is possible in the experimental investigation in question [14] that at certain instants of constriction of the filament the electron and hole concentration was sufficient for the formation of a negative temperature state. So far, however, it has not been possible to obtain evidence of the existence of negative temperature in such a method of excitation in terms of the constriction of the emission line or the direction or radiation.

3. The Generation of Negative Temperatures
in Transitions to Impurity Levels

The arguments advanced in §2 regarding the generation of negative temperatures with respect to interband transitions are equally valid with respect to transition between the conduction band and acceptor impurity levels situated near the valence band, or to transitions from donor impurity levels situated near the bottom of the conduction band into the valence band, as long as the relaxation time with the corresponding nearby bands is small in comparison with the lifetime τ_c. In this case, the condition (14) for the formation of negative temperatures may be written as follows:

$$\mu_e + \mu_h \geqslant \Delta_1, \tag{15}$$

where Δ_1 is the distance from the impurity level to the edge of the band from which the transition to this level is being considered.

Somewhat different conditions, particularly in the methods of excitation, are obtained in the generation of negative temperature states in transitions from a band to impurity levels situated near this band, i.e., when the radiation of fairly long-wave quanta takes place. For example, an antimony impurity in germanium forms a donor level situated in the forbidden band at a distance of 0.01 eV from the bottom of the conduction band, corresponding to transition with a wavelength of about 120 μ. In this case, given relatively low electric field strengths, cumulative ionization of electron-filled donor levels takes place at a sufficiently low crystal temperature (normally the temperature of liquid helium). Consequently, the first stage, ionization of the impurities in the presence of shallow donor levels, is realized fairly easily. For example, in the case of the antimony impurity in germanium, as shown in [32, 33], almost total ionization of the impurities is attained at a lattice temperature of 4°K in fields up to 20 $W \cdot cm^{-1}$.

However, as in the case of interband transitions, the carriers in a field are distributed within a wide energy interval of the corresponding band and do not form negative temperature states with respect to the impurity levels. The generation of negative temperatures in this case is possible only if the carriers inside the corresponding bands occupy a narrow energy belt ~kT near the edges of the bands upon instantaneous removal of the field, but still are not able to recombine in the impurity, i.e., it is once again necessary to meet the condition $t_S \ll \tau_C$. The carrier recombination time τ_C is determined by the recombination cross section σ_C and the number of ionized impurities, which is equal to the number of carriers n in the band and their velocity v:

$$\tau_c^{-1} = v \sigma_c n. \tag{16}$$

Inasmuch as the formation of negative temperatures requires relatively high carrier concentrations, n, clearly the largest contribution to the nonradiative recombination cross section will be provided by triple recombination processes, which are calculated by the author [34] in Appendix 1. An important consideration is the fact that the triple recombination cross section σ_C itself depends linearly on the carrier concentration n, hence τ_C decreases as the square of the nonequilibrium carrier density. In [35] the problem of shallow impurity levels is solved for typical semiconductors which, as a rule, produce a hydrogen-like spectrum whose calculation requires allowance for the dielectric constant of the semiconductor and the effective mass of the carrier. Lax [36] has shown that the capture cross section at such levels is very large, especially at low temperatures, when capture occurs at shallow levels (giant orbits) with subsequent relaxation inside the impurity at the ground level. If this fact is taken into consideration, even at a nonequilibrium carrier concentration of approximately 10^{14} cm^{-3}, the triple recombination cross section $\sigma_C \approx 5 \cdot 10^{-13}$ cm^2 in the case of the antimony impurity in germanium, consistent with [37]. As shown in Chapter V, the slowing-down time t_S is determined from

$$t_s \sim \frac{2}{a_0 (kT)^{1/2}}, \qquad a_0 = \frac{8eu^2}{3\sqrt{\pi}\, w(T)(kT)^{3/2}}, \tag{17}$$

where u is the velocity of sound in the crystal and w is the mobility in scattering by acoustic vibrations of the lattice for the temperature T.

The condition $t_s \ll \tau_c$ places a restriction on the number of ionized impurities n:

$$n^2 \ll \frac{\sqrt{m}\; a_0 n_0}{\sigma_c(n_0)}, \tag{18}$$

where n_0 is the concentration for which the value of $\sigma_c(n_0)$ is known. In the case of antimony in germanium, the condition (18) places a restriction on $n \ll 3 \cdot 10^{14}$ cm^{-3}. The negative temperature state occurs after a period of the order t_s from the instantaneous removal of the electric field and exists for a period of the order τ_c. The number of carriers n' left in the band after a period of the order t_s from the disconnection of the field is approximately determined from the relation

$$n' \approx \frac{n}{t_s} \left(\frac{1}{t_s} + \frac{1}{\tau_c} \right)^{-1} \approx n \left(1 - \frac{t_s}{\tau_c} \right), \tag{19}$$

the number of electrons at the impurity levels is equal to $N - n'$, and the corresponding population of the impurities is

$$f_i = \frac{N - n'}{N}, \tag{20}$$

where n' is the concentration of impurities in the semiconductor. If $f_i \ll 1$, degeneracy of the carriers in the band is not required for the formation of a negative temperature state. In this case, the carrier population of the band f_b may be roughly estimated according to the formula

$$\bar{f}_b \approx \frac{n'}{n_{eff}}, \tag{21}$$

where

$$n_{eff} = 2 \left(\frac{m^* kT}{2\pi\hbar^2} \right)^{3/2};$$

m* is the effective mass of the carrier.

Taking Eqs. (19)-(21) into account, we obtain from the condition (10) for formation of a negative temperature state, i.e., $\overline{f_b} > f_i$, the condition for the temperature of the sample T and the number of ionized impurities for a specific semiconductor and given impurities therein.

For $f_i \sim 1$, the temperature of the sample must be chosen so that degeneracy of the current carriers will arise in the band. In this case, the temperature must satisfy the condition

$$T < \frac{(\pi\hbar)^2}{2m^* k} \left(\frac{3n'}{\pi} \right)^{2/3}. \tag{22}$$

In the case of antimony in germanium, bearing (18) in mind, estimation according to Eq. (22) gives T < 1°K, where, according to (16), the lifetime $\tau_c \approx 0.5 \cdot 10^{-8}$ sec.

We note that fulfillment of the self-excitation conditions required for generation (see §2 of Chapter II) imposes a limitation on the minimum number of nonequilibrium carriers. It follows from the above estimates that in germanium, for example, or in other semiconductors with impurities yielding levels near the corresponding bands, the realization of a negative temperature state between these levels and the nearest band, although possible in principle, incurs considerable difficulties due to the need for working at exceedingly low temperatures under conditions of rapid field removal (front shorter than 10^{-9} sec).

Consequently, the success of the method of obtaining negative temperature states in semiconductors by excitation with electric field pulses depends by and large on the ability to produce very short pulses with a large current density. In this case, it is possible to avoid overheating of the semiconductor and to reduce considerably the Joule heat losses during the passage of current. In fact, the energy given off in Joule heat per cm^{-3} during the period of one pulse is equal to

$$E_J \approx \frac{e^2 F_i^2 n \Delta t}{m},$$

where all the notation and variables are taken from Eq. (10). On the other hand, the maximum radiation energy (in cm^{-3}) that can be acquired during one pulse from n pairs is equal to

$$E_{\text{rad}} = n\hbar\omega.$$

Taking the ratio of the radiated energy to the energy lost in Joule heat, we obtain the expression

$$\frac{E_{\text{rad}}}{E_J} = \frac{\hbar\omega}{\hbar\omega_0} \frac{\tau_{\text{op}}}{\Delta t}.$$

It follows from an analysis of this expression that even for $\Delta t \sim 10^{-8}$ sec ($\tau \sim 10^{-12}$ sec, $\hbar\omega/\hbar\omega_0 \sim 10^2$ the energy going for Joule heat is still 100 times the radiation energy. As Δt tends to τ_{op}, we greatly reduce the Joule heat losses but, clearly, with this method of excitation we cannot obtain a high efficiency, due to the losses associated with the necessary transfer of a part of the energy of the ionizing particle into kinetic energy of the generated electron and hole (this problem is analyzed in detail in §2 of Chapter IV).

CHAPTER II

THE GENERATION OF NEGATIVE TEMPERATURE
BY INJECTION THROUGH A p−n JUNCTION

1. Condition for the Onset of Negative Temperature in the Case of Injection through a p − n Junction

Another method for obtaining states with negative temperature by the excitation of a semiconductor with an electric current was proposed in [5] and has been realized in certain laboratories of the USA and at FIAN SSSR [18-20]. In this method, nonequilibrium current carriers are injected through a p−n junction of degenerate semiconductors by the application of an external voltage to the sample, creating a current through the p−n junction in the forward direction.

As opposed to the preceding method, it is not required in this mode of excitation to have high field strengths (only a few volts per centimeter). This means that the carriers are not heated on the part of the electric field. Simultaneously with relatively low Joule heat losses, a large part of the energy is liberated in radiative recombination. The realization of negative temperature states with injection through a p−n junction is feasible in the continuous mode, provided only that there is no appreciable heating of the crystal lattice. Consequently, as shown later, the efficiency of a laser utilizing injection through a p−n junction can be made to approach 100%.

Upon application of a voltage to a p—n junction in the forward direction, the concentration of minority current carriers increases in a layer near the p—n junction with a thickness on the order of the diffusion length, due to a decrease of the potential barrier formed by the space charge in series with the junction.

With complete removal of the potential barrier by an external field, the concentration of these carriers becomes equal in order of magnitude to their concentration in the part of the crystal where they are majority carriers, i.e., it may be assumed that the Fermi quasi-levels in the transition region coincide with the Fermi levels in the conduction and valence parts. It then follows from the condition (14) that the semiconductor must be degenerate in at least one part and that the minimum value of the external voltage for which a negative temperature state arises is equal to

$$\varphi_{min} = \frac{\Delta}{e} \, . \tag{23}$$

The condition (23) applies to the case of negative temperature with respect to interband transition. If a transition to an impurity level is used, Δ must be replaced in (23) by Δ_1 according to the condition (15).

In the case of a large impurity concentration, the mean free path of the carriers is much smaller than the diffusion length. Moreover, at p—n junctions of strongly degenerate semiconductors the negative temperature state sets in before there is any chance of completely eliminating the potential barrier. These considerations permit the diffusion theory of current through a p—n junction to be used for obtaining crude quantitative estimates although, to be sure, only a solution of the equations for large current densities will give the exact result. The current density $\mathcal{J}$ (for example, the electron component) is equal in order of magnitude to [38]

$$\mathcal{J} \approx \left(\frac{eDn_p}{L_e} \right) \exp \left(\frac{e\varphi}{kT} \right) , \tag{24}$$

where D is the diffusion coefficient, L_e is the diffusion length for electrons, n_p is the equilibrium electron density in the valence (p—) part of the semiconductor.

An analysis of Eq. (24) shows that the current density decreases with increasing degeneracy and with decreasing sample temperature. This makes it possible to obtain states with negative temperature in a steady-state regime. However, due to the presence of various mechanisms for the absorption and scattering of radiation in a semiconductor, a negative absorption coefficient occurs for fairly high nonequilibrium concentrations, when the gain exceeds the sum of all the absorption coefficients (this problem is studied in further detail in §2 of Chapter III). Therefore, even though the realization of negative temperature states by the method of injection through a p—n junction is feasible at relatively low current densities, the operation of the oscillator requires a higher value of the current density.

It is possible in principle to obtain negative temperature states by the injection of large concentrations of electrons and holes into a narrow region of a pure semiconductor (the width of the region being of the order of the diffusion length), with strongly doped n- and p-parts contiguous to this region. In this case, the nonequilibrium carriers would have large lifetimes, which would permit a reduction in the current, while the doped n- and p-parts would create a natural waveguide for radiation generated inside the pure semiconductor.

2. Self-Excitation Conditions and Efficiency of a Semiconductor Laser Using the Method of Injection through a p—n Junction

As already noted, the negative temperature state can be used for the amplification and generation of electromagnetic radiation, if the system as a whole has a negative absorption

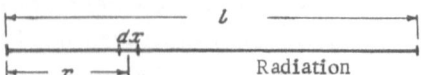

Fig. 3. Diagram showing the amplification of radiation inside the active
substance of a laser.

coefficient for the relevant phonons. This problem
will be discussed in detail in connection with indirect
transitions, where it is particularly important, in that
the resulting gain in this case is rather small. If, on
the other hand, we consider only direct transitions,
i.e., the recombination of an electron with a hole without the participation of a phonon, the problem of ampli-
fication by the system is solved fairly simply. In fact, the main absorption process in a system
for which the negative temperature condition (14) or (15) is satisfied is absorption by free carriers.

Clearly, with a strong inverted population, as occurs in practice for energy levels situated between the Fermi quasi-levels, the gain is equal to the absorption coefficient for the
given photons for a completely free conduction band and a filled valence band (or impurity
levels — see Appendix II). With the fulfillment of the condition (14) or (15), therefore, the requirement for the presence of a negative absorption coefficient for photons with energy $\hbar\omega$ satisfying the condition

$$\Delta \leqslant \hbar\omega < \mu_e + \mu_h, \tag{25}$$

means that the absorption coefficient for interband transitions at the frequency $\omega - \alpha_\omega$ must be
greater than the sum of all the rest of the absorption and scattering coefficients of the semiconductor. As shown in [39], for example, for such a semiconductor as GaAs, in which direct
transition is effected, the given condition is satisfied with a large margin. It is evident that
this also applies to all other semiconductors in the case of direct transitions (for example, to
compounds of the groups III and V).

However, the presence of a positive gain for the system still does not imply that the system is capable of functioning as an oscillator, generating monochromatic radiation. Only with
feedback in the system and with fulfillment of the self-excitation condition is the generation
mode realized [40].

In an example referring to the case of light propagation inside a semiconductor in which
the conditions (14) and (15) are satisfied, we show how the process of wave amplification and
narrowing of the emission line originates and how the system goes over into the generation
mode. For simplicity, we consider radiation propagating perpendicularly to the faces of the
semiconductor, for which the separation between the faces is equal to l (Fig. 3), the reflection
coefficient at the faces is R. Let i_0 be the intensity of spontaneous emission in the given direction from a unit length of the layer. Then the intensity of the amplified radiation emitted by a
layer of thickness dx and traversing the distance $(l - x)$ to the face is equal to

$$dJ = i_0 e^{(\alpha - \varkappa)(l-x)} dx, \tag{26}$$

where α is the gain and $\varkappa$ is the absorption coefficient.

The relation (26) may be used for any frequency ω at which amplification occurs, but with
its own unique value of the coefficients $\alpha(\omega)$ and $\varkappa(\omega)$. From the radiation surface attained, a
portion equal to $(1 - R)$ emerges outward, another part, equal to R, is reflected in the opposite
direction, amplified, reaches the opposite face, is partially reflected from it, and, being amplified, returns again to the surface in question. For this part of the radiation, the entire process
is repeated.

Consequently, allowance for all such processes gives the sum of a geometric progression:

$$dJ = i_0 dx e^{(\alpha - \varkappa) l} (1 - R)\{1 + R^2 e^{2(\alpha - \varkappa) l} + [R^2 e^{2(\alpha - \varkappa) l}]^2 + \ldots\}. \tag{27}$$

As a result of summation of the geometric progression with $R^2 e^{2(\alpha - \varkappa)l} < 1$, we find

$$dJ = i_0 dx e^{(\alpha - \varkappa)(l-x)} \frac{1 - R}{1 - R^2 e^{2(\alpha - \varkappa)l}} .$$ (28)

Taking into account also the part of the radiation traveling originally in the other direction, we obtain

$$dJ = i_0 e^{(\alpha - \varkappa)x} e^{(\alpha - \varkappa)l} dx \frac{R(1 - R)}{1 - R^2 e^{2(\alpha - \varkappa)l}} .$$ (29)

Integrating over x from 0 to l and summing, we have

$$\frac{J}{J_0} = \frac{e^{(\alpha - \varkappa)l} - 1}{(\alpha - \varkappa)l} \frac{1}{1 - R e^{(\alpha - \varkappa)l}} ,$$ (30)

where J_0 denotes the quantity $(1 - R)i_0 l$. Equation (30) reflects the main processes that occur in the system: amplification, narrowing of the emission line, transition to the generation (oscillation) mode when the denominator goes to zero. Consequently, generation commences on fulfillment of an inequality called the self-excitation condition:

$$R e^{(\alpha - \varkappa)l} \geqslant 1.$$ (31)

Near the maximum value, the dependence of the coefficient $[\alpha(\omega) - \varkappa(\omega)]$ on the frequency may be represented as a quadratic function of the difference $(\omega - \omega_0)$. In this case, on approaching the self-excitation condition, the emission line will narrow with increasing intensity, such that

$$\delta\omega \sim [1 - R e^{(\alpha_0 - \varkappa_0)l}]^{1/2}.$$ (32)

It follows from an analysis of (32) that when the self-excitation condition is fulfilled, the intensity increases without bound at the frequency ω_0, while the width of the line tends to zero. However, the inferences stemming from Eq. (32) are not applicable to the generation mode if the radiation density is so large that saturation must be accounted for in the coefficient $\alpha(\omega)$ [41]. Therefore, the foregoing conclusions only permit the self-excitation conditions of an oscillator to be determined and the narrowing of the emission line to be explained in the absence of saturation effects. The above calculations involved the gain $\alpha(\omega)$, which is calculated in Appendix II for the case of direct interband transition in a semiconductor. Appendix II implies that $\alpha(\omega)$ changes sign in the region where the factor $[f_e(\varepsilon) + f_h(\hbar\omega - \varepsilon) - 1]$ changes sign [the energy is measured from bottom to top in the conduction band for electrons, from top to bottom for holes, i.e., $\varepsilon_h = m_h(\hbar\omega - \Delta)/(m_e + m_h)$]. However, the stipulation of this factor being positive is the condition for the presence of a negative temperature state.

Consequently, in the presence of a negative temperature state, $\alpha(\omega)$ at first increases with ω as $(\hbar\omega - \Delta)^{1/2}$, reaches a maximum, then decreases due to the factor $[f_e(\varepsilon) + f_h(\hbar\omega - \varepsilon) - 1]$ and passes through zero into the negative region, where we have ordinary absorption. Due to the large value of the transition matrix element in semiconductors, the gain varies markedly with variation of the quantity $\mu_e + \mu_h$, whence it follows that upon fulfillment of the self-excitation condition, $\mu_e + \mu_h$ has a value near $\hbar\omega$. The coefficient $\alpha(\omega)$ may therefore be expanded in powers of $\mu_e + \mu_h - \hbar\omega$ and the self-excitation condition written in the form

$$\mu_e + \mu_h = \hbar\omega + \left(\varkappa - \frac{\ln R}{l}\right)\left[\frac{\partial\alpha}{\partial(\mu_e + \mu_h)}\right]^{-1}.$$ (33)

All variables on the right side of the equation are evaluated at $\omega = \omega_0$, where $\alpha(\omega_0)$ is a maximum.

Due to the saturation effect, in the generation (oscillation) mode the maximum energy separation of the Fermi quasi-levels μ_e and μ_h has a value near that given by Eq. (33) and, clearly, depends only slightly on the generated power. This makes it possible to estimate the efficiency η of a semiconductor laser for injection through a p—n junction. For this, we investigate the process of a constant current passing through an inhomogeneous semiconductor, invoking the argumentation given in Landau and Lifshits (Electrodynamics of Continuous Media) [42], recognizing the fact that conduction is realized in our case by electrons, as well as holes. For convenience in the ensuing discussion, we interpret the values of μ_e and μ_h as the sum $e\varphi + \mu_0$, where μ_0 is the chemical potential of the electrons (or holes) at the potential $\varphi = 0$. Consequently, μ_e and μ_h are now interpreted as the chemical potentials for electrons and holes, respectively, which, of course, are functions of the coordinates in the absence of thermodynamic equilibrium.

The energy flux density **q** transported in the semiconductor is equal to

$$\mathbf{q} = \frac{1}{e}\left(\mu_e \mathbf{j}_e - \mu_h \tilde{\mathbf{j}}_h\right) + \mathbf{q}_r\left(\mathbf{j}_e, \mathbf{j}_h, T\right), \tag{34}$$

where $\mathbf{j}_e$ and $\mathbf{j}_h$ are the electron and hole components of the current density, and $\mathbf{q}_T$ describes the thermoelectric effects, which we will disregard (the admitting contacts are identical and exist at the same temperature). The total current density **j** is constant and equal to

$$\mathbf{j} = \mathbf{j}_e + \mathbf{j}_h, \tag{35}$$

where $\operatorname{div}\mathbf{j} = 0$.

We next proceed with a discussion of the one-dimensional case, when only the x coordinate is significant (i.e., the coordinate axis perpendicular to the plane of the p—n junction). All the values of the variables are referred to unit area, whence the total variables are readily derived by simple multiplication by the cross section of the p—n junction. We set the origin of the x-coordinate axis in the center of the region where $\mu_e + \mu_h = 0$ with current present (points x_1 and x_2). The quantity of heat† released per second in the region between the points x_1 and x_2, per unit area, is equal to

$$jV = \frac{1}{e}\int_{x_1}^{x_2}\frac{\partial}{\partial x}\left(\mu_e j_e - \mu_h j_h\right)dx. \tag{36}$$

Differentiating and making use of Eq. (35), we obtain from (36)

$$jV = \frac{1}{e}\int_{x_1}^{x_2}\left[\frac{1}{2}j\frac{\partial}{\partial x}(\mu_e - \mu_h) + \frac{1}{2}j_e\frac{\partial}{\partial x}(\mu_e + \mu_h) - \frac{1}{2}j_h\frac{\partial}{\partial x}(\mu_h + \mu_e) + (\mu_e + \mu_h)\frac{\partial j_e}{\partial x}\right]dx. \tag{37}$$

Clearly, the integral of the first term in the bracketed expression yields jV, contracted with the term on the left side, so that we obtain

$$\int_{x_1}^{x_2}(\mu_e + \mu_h)\frac{\partial j_e}{\partial x}dx = -\int_{x_1}^{x_2}\frac{1}{2}(j_e - j_h)\frac{\partial}{\partial x}(\mu_e + \mu_h)dx. \tag{38}$$

It is reasonable that the energy going for radiation should be contained only in the term on the left side, since this radiation is accompanied by the recombination of minority current carriers. However, this term also includes the spontaneous radiation and the heat released in non-

---

† Here the released thermal energy includes both incoherent and coherent radiation.

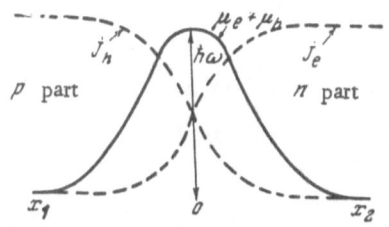

Fig. 4. Distribution of the electron and hole current densities and behavior of the Fermi quasi-levels in a p−n junction at high injection levels.

radiative recombination. We postulate that once the threshold value of the current required for the initiation of generation has been reached, all the excess energy entering into the generation region is converted into energy of coherent radiation of the oscillator. Such a postulate corresponds to the oscillator functioning in the saturation regime, where $\mu_e + \mu_h$ is near the value at the self-excitation threshold. Then the generated power is equal to

$$U \approx \frac{1}{e} \int_{x_1}^{x_2} (\mu_e + \mu_h) \frac{\partial}{\partial x} (j_e - j_{0e}) \, dx,$$

where j_{0e} and j_{0h} are the threshold values of the current density at which the generation mode is initiated.

Making use of Eq. (38), we obtain

$$U = -\frac{1}{e} \int_{x_1}^{x_2} \frac{1}{2} (j_e - j_{0e} - j_h + j_{0h}) \frac{\partial (\mu_e + \mu_h)}{\partial x} \, dx. \tag{39}$$

The efficiency η is defined as the ratio of the generated coherent light power to the power lost in the interval from x_1 to x_2, equal to jV:

$$\eta = U / jV.$$

For a precise determination of η, it is necessary to solve the very intricate problem of the distribution of the current components and Fermi quasi-levels at high injection levels through the p−n junction. However, recognizing that the sum of the Fermi quasi-levels in the generation region depends only slightly on the current, so that $\mu_e + \mu_h$ has the same dependence on x as at the threshold current (the behavior of the sum of the quasi-levels and current density components is shown in Fig. 4 as a function of x), the efficiency may be rather well approximated by the following expression:

$$\eta = \frac{\hbar \omega (j - j_0)}{e j V} \frac{1 - e^{-\varkappa l}}{\varkappa l}. \tag{40}$$

The second factor in (40) is introduced to account for the absorption by free carriers prior to the emergence of radiation from semiconductors. An analysis of Eq. (40) shows that the efficiency of a semiconductor laser utilizing the method of carrier injection through a p−n junction, with the stipulation of operation far above threshold (the generation threshold should preferably be low) and with small ohmic losses, can turn out to be nearly unity.

CHAPTER III

STATE WITH NEGATIVE TEMPERATURE AND NEGATIVE
ABSORPTION COEFFICIENT FOR INDIRECT TRANSITIONS
IN SEMICONDUCTORS

The fundamental notion behind the method of generating negative temperature states† for indirect transitions in semiconductors, as first proposed in [2-4], is contained in the following.‡ As we know, in the recombination of an electron—hole pair with the emission of just one photon, the energy of the photon is equal in order of magnitude to the width Δ of the forbidden band, and the momentum has a very small value of the order Δ/c. In other words, such a process is tantamount to the recombination of an electron and hole, each of which is almost equal in absolute value but opposite to the directed quasi-momentum. In the band diagram, this transition corresponds to a vertical transition of an electron from the conduction band into the valence band (direct transition).

The recombination of an electron—hole pair with a large total quasi-momentum can only take place with the participation of phonons. In this case, the emitted photon has an energy of the same order as the width of the forbidden band, while the momentum of the phonon is essentially equal to the initial momentum of the pair. In the band diagram, this process corresponds to a transition of an electron from the conduction band into the valence band with substantially different values for the quasi-momentum (indirect transition) [43].

At low sample temperatures, the electrons and holes, having a kinetic energy of the order kT, occupy the lower edges of the corresponding bands. In the event that the structure of the energy bands of the semiconductor is such that the minimum of the conduction band and the maximum of the valence band have the same quasi-momentum, direct transitions will prevail in recombination, otherwise indirect transitions will prevail. In semiconductors such as germanium and silicon, the second situation is realized, and radiative recombination of an electron and hole existing at the edges of the bands corresponds to indirect transitions [43, 44].

In this chapter, it will be shown that a process corresponding to the emission of radiation with the longest wavelength (indirect transition with simultaneous emission of a photon and phonon) in semiconductors with a band structure like germanium and silicon may be used for the creation of negative temperature states at relatively low nonequilibrium current carrier concentrations, if the temperature of the sample is low.

The fundamental concept of the method lies in the fact that when phonons are absent in the crystal (low temperatures), the converse of the process of emission of the longest-wave photons, i.e., simultaneous absorption of a photon and phonon, will not occur. On the other hand, the probability of induced emission of photons will be finite, even as $T \to 0$. It is to be expected, therefore, that at very low temperatures of the sample, if the slowing-down time (see §2 of Chapter V) is less than the lifetime, even a small increase in the number of carriers over the

---

† Here we understand the negative temperature state not to mean an inverted population, as in the case of direct transitions, but a state yielding amplification of external radiation with regard for the direct and reverse processes.

‡ For the most part, the present chapter is a reiteration of material presented in the dissertation of Krokhin[55], who collaborated with N. G. Basov and the present author in the development of this method. An important addendum to the problems treated by Krokhin is the allowance for large concentrations of nonequilibrium carriers, whereupon it is necessary to include degeneracy.

equilibrium count is sufficient for the probability of induced emission of photons to exceed the probability of their absorption in the converse process.

The system we are considering is analogous to a three-level energy diagram, where the first level corresponds to the ground state of the crystal, when phonons and electron−hole pairs are totally absent, the second level corresponds to the presence of a phonon with energy $\mathscr{E}$ in the crystal, and the third level corresponds to the presence of an electron−hole pair in the crystal with an energy near the minimum width of the forbidden band Δ. The emission of a photon with energy $\hbar\omega$ represents in this diagram a transfer of the system from the third to the second level, with the appearance of a phonon with energy $\mathscr{E}$ in the lattice following the event of emission. The converse process, corresponding to transition from the second to the third level with the absorption of a photon, is proportional to the degree of filling of the second level, i.e., the probability of a phonon being present in the lattice. This probability is equal to $e^{-\mathscr{E}/kT}$ at very low temperatures and diminishes as $T \to 0$.

The degree of filling of the third level, on the other hand, is $f_e f_h$, where f_e is the probability of an electron, f_h is the probability of a hole being present in the semiconductor (for an intrinsic semiconductor, $f_e = f_h$). In thermodynamic equilibrium, $f_e f_h = e^{-\Delta/kT}$. Inasmuch as the filling of the second level is very small, for the onset of states in which the probability of induced emission exceeds the probability of absorption, a relatively small increase in the degree of filling of the third level is sufficient by comparison with the equilibrium population determined from the condition $f_e f_h > \exp(-\mathscr{E}/kT)$. If we invoke the concept of the "effective temperature" T_e, which characterizes the nonequilibrium excited third level, the preceding inequality gives the condition for the effective temperature T_e.

Consequently, on fulfillment of the above condition, the degree of filling of the third level exceeds that of the second, thus realizing a negative temperature state. In the present chapter, we give some numerical calculations, which show that it is necessary, in order to obtain a negative temperature state in germanium at a temperature $T = 10°K$, to have a nonequilibrium current carrier density on the order of 10^{11} cm^{-3}. It is important to emphasize that in this case a negative temperature state arises with respect to the collective excitation levels produced in the crystal both by vibrations of the lattice and by electrons and holes, but not with respect to the electron states in the conduction band and valence band, as was the case with direct transitions.

Our subsequent analysis in §2 of this chapter shows that the presence of a variety of mechanisms for the absorption of radiation, primarily free-carrier absorption, can cancel out the effect of light amplification in indirect transitions. However, at higher values of the current-carrier concentrations, it is possible to obtain negative absorption-coefficient states. This is related to the dissimilar carrier-concentration dependence of the probability of induced emission in indirect transitions, which is proportional within well-defined limits to the concentration squared, then goes over to the $4/3$ power, and the probability of absorption, which is also linear with respect to the carrier concentration within certain limits.

Another possible mechanism that can be used to obtain negative temperature states is the indirect recombination of an electron and hole from the exciton state [45-48]. An exciton may be regarded in a certain sense as an electron and hole existing in the bound state. The energy of the exciton bond is attributable to the forces of electrostatic attraction between the electron and hole, attenuated by the polarization of the crystal, and may be calculated in the first approximation on the basis of a hydrogen-like model. In semiconductors such as germanium and silicon, the bond has a magnitude of the order of 0.01 eV.

In §1 of the present chapter we examine the conditions under which a negative temperature state arises with respect to indirect transitions from exciton states. The very high probability of exciton recombination is due to the fact that an electron and hole found in the bound state are localized within a small region of the semiconductor. As shown in §3 of this chapter, this permits one to obtain a negative absorption-coefficient state at relatively low exciton densities (one or two orders smaller than the corresponding concentrations for the unbound carriers).

The shape of the emission line obtained in §2 of this chapter for indirect recombination shows that its width is of the order of several kT. With a reduction in temperature, the emission line narrows, thus facilitating the onset of a negative absorption coefficient (and, hence, conditions for the self-excitation of oscillators). This clearly promotes an increase in the radiative recombination cross section with decreasing carrier velocity. However, a limit is set on the above-indicated useful effects by possible degeneracy of the electron−hole gas, as well as by collisions with the lattice and impurities.

1. Conditions for the Formation of Negative Temperature States in Indirect Transitions

We begin by calculating the difference in the number of transitions in the system for the process of simultaneous emission of a photon and phonon in the recombination of an electron and hole, and the process of absorption of a photon and phonon with the creation of an electron−hole pair.†

The number of transitions per unit time with simultaneous emission of a photon and phonon is equal to

$$I^+ = W\,(n_r + 1)\,(n_q + 1)\,f_e^i f_h^j, \tag{41}$$

where W is the transition probability, n_r and n_q are, respectively, the number of photons and phonons in a given state (r or q), and f_e^i, f_h^j are the distribution functions for electrons and holes, respectively, in the quantum states i and j. We will assume below that the functions f_e and f_h depend only on the energy of the electron (ε_e) and hole (ε_h), and that the number of phonons n_q, which is determined by the temperature of the crystal, depends only on the phonon energy $\mathscr{E}$. The energy of the emitted photon $\hbar\omega$ is related to the phonon energy and kinetic energy of the electron and hole by the following conservation law:

$$\varepsilon_e^i + \varepsilon_h^j + \Delta - \mathscr{E} - \hbar\omega = 0, \tag{42}$$

where Δ is the minimum width of the forbidden band.‡

The number of transitions per unit time with the absorption of a photon and phonon (converse process) is equal to

$$I^- = W n_r n_q\,(1 - f_e^i)\,(1 - f_h^j). \tag{43}$$

According to the principle of detailed balancing [23], we adopt in Eqs. (41) and (43) the same probability for the direct and converse processes for transition between definite quantum states.

In the state of thermodynamic equilibrium, the number of events with emission is equal to the number of events with absorption, i.e.,

$$I^+ - I^- = 0. \tag{44}$$

---

† We do not consider many-phonon processes, because their relative probability is greatly reduced as the number of phonons emitted is increased.

‡ Here the energy of the electrons is measured upwards from the minimum of the conduction band, the energy of the holes downward from the maximum of the valence band.

If we substitute for the quantities f_e, f_h, n_q, n_r in Eqs. (41) and (43) their thermodynamic equilibrium values:

$$f_e = (e^{(\varepsilon_e - \mu_e)/kT} + 1)^{-1}, \quad f_h = (e^{\varepsilon_h - \mu_h/kT} + 1)^{-1},$$

$$n_r = (e^{\hbar\omega/kT} - 1)^{-1}, \quad n_q = (e^{g/kT} - 1)^{-1},$$

taking into account the law of conservation of energy (42) and the equality of the Fermi levels in equilibrium, then Eq. (44) is an identity. The expression (41) determining the transitions in a system with the emission of quanta represents the sum of two terms: one term describing spontaneous transitions with respect to the emission of electromagnetic quanta:

$$I_s^+ = W(n_q + 1) f_e^i f_h^j, \tag{45}$$

another term proportional to the number of photons in the system n_r and describing the induced emission of photons:

$$I_i^+ = W n_r (n_q + 1) f_e^i f_h^j. \tag{46}$$

Only the induced emission of quanta has bearing on the amplification of electromagnetic waves, since only in this case are the photons emitted by the system completely identical with the photons by which this transition is stimulated. In the state of thermodynamic equilibrium, the inequality $I_i^+ < I^-$ always holds, i.e., the absorption of quanta of a definite type and their subsequent re-emission by spontaneous transition always occurs. This, in particular, explains why the radiation emitted by heated bodies is incoherent.

In order for the system to exist in a state with negative temperature, it is essential to upset thermodynamic equilibrium, i.e., to increase the nonequilibrium concentration of electrons and holes in the sample to the extent that the number of events of induced emission will exceed the number of events with the absorption of quanta. Taking (46) and (43) into account, we obtain

$$(n_q + 1) f_e^i f_h^j - n_q (1 - f_e^i)(1 - f_h^j) > 0. \tag{47}$$

We assume that the phonons are in thermodynamic equilibrium with the crystal lattice, which has a temperature T, so that $n_q/(1 + n_q) = e^{-g/kT}$. Then Eq. (47) may be rewritten in the form

$$\frac{f(\varepsilon_e) f(\varepsilon_h)}{[1 - f(\varepsilon_e)][1 - f(\varepsilon_h)]} > \exp\left(-\frac{g}{kT}\right). \tag{48}$$

From this moment on, we drop the superscripts i and j and agree that the subscript e refers to electrons, with application also to the electron distribution function, the subscript h to holes. We note that the following relation is always valid for semiconductors in thermodynamic equilibrium:

$$\frac{f(\varepsilon_e) f(\varepsilon_h)}{[1 - f(\varepsilon_e)][1 - f(\varepsilon_h)]} = \exp\left(-\frac{\Delta + \varepsilon_e + \varepsilon_h}{kT}\right). \tag{49}$$

As noted earlier, an important characteristic of the process under discussion is the fact that with decreasing sample temperature, the creation of negative temperature states requires less-appreciable departures from the state of thermodynamic equilibrium. In particular, this is apparent from Eq. (48). Consequently, at low temperatures, the necessary nonequilibrium current carrier densities are small, and the negative temperature state occurs long before degeneracy of the electron−hole gas has a chance to set in. This simplifies Eq. (48), and the condition for the onset of a negative temperature state assumes the form

$$f\left(\varepsilon_e\right) f\left(\varepsilon_h\right) > \exp\left(-\frac{\mathscr{E}}{kT}\right). \tag{50}$$

The form of the energy distribution functions for electrons and holes is determined by the condition for the generation of electron—hole pairs by an external source, interaction of the carriers with the lattice, causing the former to be slowed down and possibly impelled into thermodynamic equilibrium with the lattice, as well as by the process of recombination.

The problem of calculating the slowing-down time (time for thermolysis with the lattice inside the corresponding band) will be taken up in Chapter V. It is shown there, in particular, that the slowing-down time is determined mainly by interaction with acoustic vibrations of the lattice and is on the order of 10^{-10} to 10^{-11} sec for semiconductors such as germanium and silicon.

The recombination rate, or lifetime of the carriers in the band, considerably exceeds the slowing-down time, amounting to an order of magnitude of 10^{-4} or 10^{-5} sec [49]. This enables one to write the electron and hole distribution functions in the form of the ordinary Fermi distribution (2), which now involves the Fermi quasi-levels for electrons and holes. Making use of this fact, we find from the condition (48) that the separation of the Fermi quasi-levels must satisfy the following inequality for the creation of negative temperatures for the emission of photons $\hbar\omega$:

$$\mu_e + \mu_h > \hbar\omega. \tag{51}$$

The inequality (51) is valid either without regard for degeneracy of the nonequilibrium carriers in the bands or with such degeneracy present. It follows from the condition (51) that the realization of negative temperature states is easiest for the longest-wave photons, whose energy is determined from the condition $\hbar\omega = \Delta - \mathscr{E}$. Consequently, the minimum separation of the Fermi quasi-levels must satisfy the condition

$$\mu_e + \mu_h > \Delta - \mathscr{E}. \tag{52}$$

According to (52), the Fermi quasi-levels may be situated in the forbidden band, such that the sum of their energy distances from the edges of the corresponding bands must be less than the energy $\mathscr{E}$ of the phonon participating in an indirect transition, i.e., degeneracy of the nonequilibrium carriers is not mandatory.

Consequently, assuming the Boltzmann function for the distribution of electrons and holes inside the corresponding bands, multiplying the inequality (50) by the level density $\rho(\varepsilon) = 4\pi(2\pi\hbar)^{-3}(2m)^{3/2}\varepsilon^{1/2}$, and integrating over the energy interval occupied by the electrons and holes ($\sim kT$), we obtain expressions for the concentrations of electrons n and holes p necessary for realization of the negative temperature state:

$$np > n_{\text{ef}}p_{\text{ef}}\, e^{-\mathscr{E}/kT}, \tag{53}$$

where

$$n_{\text{ef}} = 2\left(\frac{m_e kT}{2\pi\hbar^2}\right)^{3/2} \approx 4.8\cdot10^{15}\left(\frac{m_e}{m}\, T\right)^{3/2},$$

$$p_{\text{ef}} = 2\left(\frac{m_h kT}{2\pi\hbar^2}\right)^{3/2},$$

m_e is the effective mass of the electron, and m_h is the effective mass of the hole (we point out that the relation $n_0 p_0 = n_{\text{ef}}p_{\text{ef}}e^{-\Delta/kT}$ holds for the equilibrium values $n = n_0$, $p = p_0$).

The condition (53) for a negative temperature state is easily realized in silicon and germanium, since it requires that

$$np > 10^{22} \text{ cm}^{-6} (\mathscr{E} \sim 0.03 \text{ ev}, \ T \sim 10^\circ \text{ K}).$$

The occurrence of negative temperature states is possible also in the case of the indirect process of electron−hole recombination from the exciton state.

By analogy with Eq. (47), we write the difference between the number of transitions with induced emission and absorption of light quanta:

$$W n_r [f_{ex}^i (n_q + 1) - n_q] > 0, \tag{54}$$

where W is the probability of radiative recombination of an exciton that exists in the state i with a probability $f_{ex}^i \ll 1$. As before, we assume that the quantity f_{ex}^i depends only on the exciton energy and has the form of a Boltzmann distribution function† $f_{ex}^i = e^{(\mu_{ex} - \varepsilon_{ex})/kT}$. Then from the inequality (54) we obtain

$$f_{ex}(\varepsilon_{ex}) > \exp\left(-\frac{\mathscr{E}}{kT}\right). \tag{55}$$

Assuming that the exciton has an effective mass equal to m_{ex} ($m_{ex} \approx m_e + m_h$), Eq. (55) may be averaged over an interval of the order kT of the exciton kinetic energy $\varepsilon = \varepsilon_{ex} - (\Delta - u)$, where u is the ionization energy of the exciton (the electron−hole binding energy u > 0). This gives the condition for the necessary exciton density ν:

$$\nu > \nu_{ef} e^{-\mathscr{E}/kT}, \tag{56}$$

where

$$\nu_{ef} = \int_0^\infty e^{-\mathscr{E}/kT} \rho_{ex}(\varepsilon) \, d\varepsilon, \quad \rho_{ex} = \frac{2\pi (2m_{ex})^{3/2} \varepsilon^{1/2}}{(2\pi\hbar)^3}, \quad \nu_{ef} = \left(\frac{m_{ex} kT}{2\pi\hbar^2}\right)^{3/2}.$$

The exciton density in the crystal is conveniently expressed in terms of the concentrations of electrons and holes found in equilibrium with the excitons. In this case, since the process of exciton formation by a free electron and hole takes place with the emission of a phonon, we can write

$$\mu_{ex} = \mu_e + \mu_h, \tag{57}$$

where μ_{ex} is the chemical potential of the exciton (the chemical potential of the phonons is equal to zero).

In the state of thermodynamic equilibrium, $\mu_e = -\mu_h$, hence $\mu_{ex} = 0$, yielding the following expression for the thermodynamic-equilibrium concentration of excitons ν_0:

$$\nu_0 = \nu_{ef} \exp\left(-\frac{\Delta - u}{kT}\right). \tag{58}$$

It is also possible from Eq. (57) to obtain the relation between the exciton density and density of free current carriers existing in equilibrium with one another. Inasmuch as

---

† The excitons formed by an electron and hole have an integer spin, hence they are subject to Bose−Einstein statistics. Consequently, an arbitrarily large number of excitons can exist in the same quantum state.

$$e^{\mu_{ex}/kT} = \frac{\nu}{\nu_{ef}} e^{(\Delta-u)/kT} \tag{59}$$

even in the case when there is no degeneracy of the free carriers,

$$e^{\mu_e/kT} = \frac{n}{n_{ef}} \; ; \quad e^{\mu_h/kT} = \frac{p}{p_{ef}} e^{\Delta/kT}. \tag{60}$$

Rewriting Eq. (59) in the form

$$e^{\mu_{ex}/kT} = e^{(\mu_e+\mu_h)/kT}, \tag{61}$$

we obtain

$$\frac{\nu}{np} = \frac{\nu_{ef}}{n_{ef}p_{ef}} e^{u/kT}. \tag{62}$$

Equation (62) is written on the assumption that equilibrium is maintained between the electrons, holes, and excitons, even though the concentrations of current carriers and excitons are themselves thermodynamically not in equilibrium and are not determined by the temperature of the crystal. In this case, the condition (57) is valid, but $\mu_e \neq -\mu_h$ and, consequently, $\mu_{ex} \neq 0$. This hypothesis is clearly true, provided only that the time of formation of the exciton by free electrons and holes is less than the recombination time of the exciton or free carriers. The exciton formation time, requiring the emission of a phonon, agrees in order of magnitude with the slowing-down time calculated in §2 of Chapter V, which is normally much less than the lifetime. If, however, this condition is not met, Eq. (62) is invalid, and in this case the formation of negative temperature states becomes unfeasible for exciton states.

The condition (56) for the occurrence of negative temperature states for exciton states may be rewritten as a condition for the necessary concentrations of free current carriers. Substituting the value of ν from Eq. (62), we obtain the inequality

$$np > n_{ef}p_{ef} e^{-(\mathscr{E}+u)/kT}. \tag{63}$$

The expression (63) differs from the previously determined inequality (53) by the factor $e^{-u/kT}$, which greatly diminishes the nonequilibrium carrier concentration at a sample temperature $kT \ll u$.

Consequently, it follows from the derived equations that a negative temperature state for indirect transitions in semiconductors with properties similar to germanium and silicon sets in at low nonequilibrium carrier concentrations (on the order of 10^{11} cm^{-3} at a sample temperature of about 10°K). Such values of the concentrations are fully realizable experimentally, for example, as the result of irradiating a sample with light whose quanta have an energy exceeding the width of the forbidden band. The intensity of the exciting light required for this is easily estimated if the lifetime of the nonequilibrium carriers and magnitude of the absorption coefficient are known. If the absorption coefficient is of the order 10^3 cm^{-1} (so that the sample thickness must be equal to 10^{-3} cm), for a lifetime of the order 10^{-5} sec the light intensity needed for the creation of 10^{11} cm^{-3} nonequilibrium carriers must be equal to 10^{13} quanta·cm^{-2}·sec^{-1}. The experimental generation of such an intensity presents no particular difficulty.

2. Negative Absorption Coefficient in Indirect Transitions of Free Carriers

As mentioned earlier [41], the condition for the onset of a negative absorption coefficient in a medium existing in the negative temperature state with respect to the type of transition under consideration says that the emission induced by a light quantum propagating in the medium must exceed the absorption of this quantum for transitions between other levels of the system, with respect to which the temperature is positive.

In the given case of indirect interband transitions, the absorption of radiation by free carriers — electrons and holes — is just such an unavoidable phenomenon. According to the laws of conservation of energy and momentum, an electron (or hole) in transition between different levels of the free state cannot absorb only one photon, hence the absorption of photons by free carriers occurs with the simultaneous emission or absorption of phonons. Since absorption by free carriers comprises a transition within a single band, while radiative recombination corresponds to an interband transition, it is to be expected (despite the fact that both processes involve the participation of phonons) that the matrix element in the case of interband recombination, in some semiconductors at least, will be greater than the matrix element for absorption by a free carrier.

The absorption of photons by free carriers in semiconductors has been studied in a great many experimental and theoretical investigations [50-52]. In [52], the cross section for absorption of a photon by a current carrier in a semiconductor was calculated in the second approximation of perturbation theory. This process corresponds to the absorption of a photon and the simultaneous absorption or emission of a phonon.

We now derive the condition for realization of a negative absorption coefficient in a semiconductor, assuming that the photons at a certain frequency (longest-wave recombination radiation) are subjected to two processes: growth due to induced indirect interband transitions and decay due to absorption of scattering by free carriers (electrons and holes). As indicated earlier, at low temperatures the absorption in interband transitions for long-wave radiation is small, i.e., the negative temperature condition is fulfilled, hence this absorption need not be taken into account. The number of induced radiative transitions in the system described above per unit volume and unit time is equal to

$$\frac{dM}{dt} = \sum_{ij} n_r f_e^i f_h^j W_{ij}, \tag{64}$$

where W_{ij} is the probability of indirect recombination of an electron in the state i with a hole in the state j. The quantity W_{ij} is expressed in terms of the recombination cross section σ: $W_{ij} = v\sigma$, where v is the relative velocity of the electron and hole. The probability W_{ij} for semiconductors was calculated in [53] with allowance for the process of phonon emission and absorption. We are interested, of course, only in that part of W_{ij} which is related to spontaneous emission of a phonon; this is precisely what we will mean by W_{ij}.

The expression derived in [53] for W_{ij} contains the product of two unknown constants defining the matrix element for interaction of the carriers with phonons and photons in interband transitions of an electron. The principle of detailed balancing is used to relate the probability of indirect recombination W_{ij} to the absorption coefficient for light [54] whose quanta have an energy corresponding to the long-wave fundamental absorption edge of a semiconductor. This makes it possible to determine completely the product of the unknown constants and thus to find the value of W_{ij}.

We will make use below of the recombination probability averaged over an energy interval of the order kT: $<W_{ij}>$. Omitting the factor n_r, we rewrite Eq. (64) in averaged form:

$$\frac{dM}{dt} = <W_{ij}> (n_0 + dn)(p_0 + dp). \tag{65}$$

Equation (65) is applicable to the case of small deviations of the concentration from its equilibrium value, which is equal to n_0 and p_0. After computation of the thermodynamic-equilibrium carrier generation rate from Eq. (65), we obtain

$$\frac{dM}{dt} = (n_0 + p_0)\, dn < W_{ij} > (dn = dp). \tag{66}$$

It follows from (66) that the decay rate of the nonequilibrium carrier density is determined by the time constant

$$\tau_R = \frac{1}{(n_0 + p_0)\,\langle W_{ij}\rangle}\,.$$

We will have need of this equation later on for the evaluation of the matrix element for radiative indirect transition. According to [53], the principal temperature dependence of τ_R is determined by the factor

$$\tau_R \sim \frac{1}{n_0 + p_0}\,\frac{e^{\mathscr{E}/kT} - 1}{e^{\mathscr{E}/kT} + 1}$$

and for $\mathscr{E} \gg kT$ essentially depends on the temperature through the carrier concentration, since only the possibility of spontaneous emission of a phonon is possible under these circumstances. Since we have a continuous energy spectrum, we go from summation to integration in Eq. (64) and consider radiation in the interval $\Delta\omega$, so that n_r is the average number of quanta associated with the radiation oscillator ω_r, and the number of quanta N is defined as

$$N = \int_{\omega}^{\omega + \Delta\omega} n_r \rho(\omega)\, d\omega \approx n_r \rho(\omega)\, \Delta\omega, \tag{66a}$$

where $\rho(\omega) = \omega^2/\pi^2 c^3$.

We write the expression for the increase in the number of quanta N due to induced transitions:

$$\left(\frac{dN}{dt}\right)^+ = n_r \hbar \Delta\omega \int_0^{\mathscr{E} + \hbar\omega - \Delta} W_{ij}(\varepsilon_e \varepsilon_h)\, f_e(\varepsilon_e)\, f_h(\varepsilon_h)\, \rho_e(\varepsilon_e)\, \rho_h(\varepsilon_h)\, d\varepsilon_e. \tag{67}$$

We have taken into account the fact that summation in (64) must be carried out only over those states i and j of the electrons and holes that satisfy the energy conservation law

$$\varepsilon_e^i + \varepsilon_h^j + \Delta - \mathscr{E} - \hbar\omega = 0. \tag{68}$$

For the state density in Eq. (67) we have

$$\rho_e = \frac{4\pi\,(2m_e)^{3/2}\varepsilon_e^{1/2}}{(2\pi\hbar)^3}, \quad \rho_h = \frac{4\pi\,(2m_h)^{3/2}\varepsilon_h^{1/2}}{(2\pi\hbar)^3}\,.$$

In order to compute the integral in Eq. (67), it is necessary first to decide the form of the distribution function for the electrons and holes. In the preceding section we indicated that the carriers are in equilibrium with the lattice within the corresponding bands, even though the concentration does not have its equilibrium value. Consequently, the electron and hole distribution function is the Fermi function (2) with appropriate quasi-levels μ_e and μ_h, measured from the bottom of the conduction band (the positive axis is upward for electrons, downward for holes), i.e.,

$$f_e(\varepsilon_e) = \frac{1}{e^{\frac{\varepsilon_e - \mu_e}{kT}} + 1}, \quad f_h(\varepsilon_h) = \frac{1}{e^{\frac{\varepsilon_h + \Delta - \mu_h}{kT}} + 1}. \tag{69}$$

However, the integral in (67) is readily evaluated in two limiting cases with the following physical interpretation: absence of degeneracy (Boltzmann distribution) and degeneracy at a temperature of absolute zero. We will analyze the first case, when carrier degeneracy does not exist. As indicated in the preceding section, a negative temperature state will be attained if $\mu_e + \mu_h > \Delta - \mathscr{E}$. Consequently, we investigate the case when the separation of the Fermi quasi-levels is contained within the limits $\Delta > \mu_e + \mu_h > \Delta - \mathscr{E}$. If we obtain a negative absorption coefficient at concentrations conforming to this stipulation, degeneracy will not in general be required, otherwise it will be necessary to account for carrier degeneracy. Substituting the Boltzmann distribution functions into (67), we find

$$\left(\frac{dN}{dt}\right)^+ = n_r \hbar \Delta \omega \frac{(4\pi)^2 (4 m_e m_h)^{3/2}}{(2\pi\hbar)^6} \exp\left\{\frac{\mu_e + \mu_h - \hbar\omega - \mathscr{E}}{kT}\right\} \times$$

$$\times \int_0^{\hbar\omega + \mathscr{E} - \Delta} W_{ij}(\varepsilon_e, \hbar\omega + \mathscr{E} - \Delta - \varepsilon_e) \varepsilon_e^{1/2} (\hbar\omega + \mathscr{E} - \Delta - \varepsilon_e)^{1/2} d\varepsilon_e. \tag{70}$$

Inasmuch as the integral in Eq. (70) is taken over a relatively small energy interval ($\varepsilon_e \sim kT \ll \omega$), we proceed as in the derivation of Eq. (65) and take the average probability $<W_{ij}>$ outside the integral sign, replacing the variable of integration by $x = (\hbar\omega + \mathscr{E} - \Delta)/kT$. Then the integral reduces to the well-known Euler beta function:

$$B\left[(\alpha + 1), (\beta + 1)\right] = \int_0^1 x^\alpha (1-x)^\beta dx = \frac{\Gamma(\alpha + 1)\,\Gamma(\beta + 1)}{\Gamma(\alpha + \beta + 2)},$$

$$\left(\frac{dN}{dt}\right)^+ = n_r \hbar \Delta \omega \frac{<W_{ij}>(4\pi)^2 (4 m_e m_h)^{3/2} \Gamma^2(3/2)}{(2\pi\hbar)^6 \Gamma(3)} (\hbar\omega + \mathscr{E} - \Delta)^2 e^{\frac{\mu_e + \mu_h - \hbar\omega - \mathscr{E}}{kT}}$$

The quantity $e^{(\mu_e + \mu_h)/kT}$ may be expressed in terms of the current carrier densities according to Eqs. (60), after which the expression for $(dN/dt)^+$ takes the form

$$\left(\frac{dN}{dt}\right)^+ = n_r \langle W_{ij} \rangle \frac{4 n p \hbar \Delta \omega}{\pi kT} y^2 e^{-y} \frac{\Gamma^2(3/2)}{\Gamma(3)}, \tag{72}$$

where

$$y = \frac{\hbar\omega + \mathscr{E} - \Delta}{kT}. \tag{73}$$

Equation (72) may be used to solve the problem of the shape of the spontaneous emission line:

$$G(\omega) \sim \left(\frac{\hbar\omega + \mathscr{E} - \Delta}{kT}\right)^2 \exp\left(-\frac{\hbar\omega + \mathscr{E} - \Delta}{kT}\right). \tag{74}$$

Consequently, the maximum intensity of spontaneous emission corresponds to a frequency determined from the condition

$$\hbar\omega = \Delta - \mathscr{E} + 2kT.$$

We note that the averaging performed in Eq. (71) facilitates numerical estimates by means of the expression (67), where the averaging was carried out analogously, but the shape of the line given by (74) is exact only in the case when W_{ij} is independent of the relative carrier velocity in an energy interval of the order kT. In this case, the radiative recombination cross section σ is inversely proportional to the relative velocity of the electron and hole.

The shape of the line is discussed in [55] for an arbitrary power-law dependence of W_{ij} on the relative velocity, and the integral (70) is computed in the case of Coulomb attraction between the electrons and holes. Here, therefore, we will not delve into these cases, referring the interested reader instead to the cited paper.

We now turn to a consideration of processes involving the absorption of electromagnetic radiation. It is undeniable that for amplification purposes the semiconductor must be the purest possible, i.e., absorption by impurities and scattering by inhomogeneities must be absent. In the presence of a negative temperature state, absorption of the amplified radiation in connection with interband transitions is absent. However, absorption by free current carriers is theoretically unavoidable. Absorption by free carriers has been investigated in [50-52] and the absorption cross section in various semiconductors determined.

The variation of the number of quanta per unit volume and unit time associated with various absorption processes is determined by the expression

$$\left(\frac{dN}{dt}\right)^{-} = -\varkappa cN,$$ (75)

where $\varkappa$ is the coefficient of absorption of electromagnetic radiation by free current carriers and c is the velocity of light in the material.

The absorption coefficient may be expressed in terms of the cross section of absorption of quanta by holes (σ_h) and electrons (σ_e):

$$\varkappa = n\sigma_e + p\sigma_h.$$ (76)

The condition for a negative absorption coefficient yields

$$\left(\frac{dN}{dt}\right)^{+} > -\left(\frac{dN}{dt}\right)^{-}.$$ (77)

Inserting the corresponding expressions for $(dN/dt)^{+}$ and $(dN/dt)^{-}$ from Eqs. (72) and (75) into (77) and borrowing the expression for n_r from (66a), we obtain

$$\frac{c^2 \langle W_{ij}\rangle 4\pi\Gamma^2 (3/2)\, y^2 e^{-\prime\prime}\hbar np}{\omega^2\Gamma (3)\, kT\, (n\sigma_e + p\sigma_h)} > 1.$$ (78)

It is convenient to express the result in terms of the radiative wavelength in vacuum $\lambda = \sqrt{D}(2\pi c/\omega)$, where D is the dielectric constant of the semiconductor. Substituting into (78) the numerical values for $\Gamma(3) = 2$, $\Gamma(3/2) = \sqrt{\pi}/2$, we rewrite Eq. (78) in the form

$$\frac{\langle W_{ij}\rangle \hbar\lambda^2 n p y^2 e^{-\prime\prime}}{8kTD\,(n\sigma_e + p\sigma_h)} > 1.$$ (79)

The average probability $\langle W_{ij}\rangle$ may be expressed in terms of the spontaneous radiation lifetime, as implied by Eq. (67). For a pure semiconductor, n = p, and we obtain the following condition for the nonequilibrium carrier concentration:

$$n = p > \frac{8kTD\,(\sigma_e + \sigma_h)}{\hbar\lambda^2 y^2 e^{-\prime\prime}\langle W_{ij}\rangle}.$$ (80)

It follows from the resulting expressions, due to the square-law dependence of $(dN/dt)^{+}$ on the number of electrons and holes, while the absorption is proportional to the first power of the carrier concentration, that a negative absorption coefficient can be obtained for a sufficiently high concentration of nonequilibrium carriers. However, the square-law growth of the radiation intensity with current-carrier concentration is valid only in the case when there is no carrier degeneracy in the bands.

An estimate made according to Eq. (80) for germanium with $\lambda \sim 1\,\mu$, $D \sim 10$, $T \sim 4°K$, $\tau_R \sim 1$ sec, with $n_0 + p_0 \sim 2.4 \cdot 10^{13}$ cm^{-3}, if $\sigma_n = 0.3 \cdot 10^{-17}$ cm^2 and $\sigma_h = 0.15 \cdot 10^{-17}$ cm^2 [39], yields a value of the concentration $n = p > 10^{17}$ cm^{-3}. At such a concentration and temperature, however, degeneracy will occur, and the above considerations are inapplicable in the case of a negative absorption coefficient for indirect transitions in germanium.

It is important to stress the fact that the conceptual possibility of obtaining a negative absorption coefficient depends on the relation of the probabilities of the emission of electromagnetic quanta in indirect recombination of slow electrons and holes and the probability of the absorption of quanta with the transition of an electron or hole to a higher energy level within the band. Negative absorption can only be obtained in the event that the probability of recombination of slow particles is greater than the probability of the absorption of a quantum by a carrier.

The conceptual possibility of obtaining a negative absorption coefficient for any particular semiconductor utilizing indirect transitions may be evaluated from the following considerations, presented in [39].

The absorption coefficient $\varkappa$ for indirect transitions may be expressed by the formula [56]

$$\varkappa = B_1 n_q (\hbar\omega + \mathscr{E}_q - \Delta)^2 + B_2 (n_q + 1)(\hbar\omega - \mathscr{E}_q - \Delta)^2, \tag{81}$$

where B_1 and B_2 are coefficients describing the processes of absorption and emission of a phonon, respectively, for the absorption of a quantum $\hbar\omega$. It is clear that Eq. (81) is easily derived from the expression (67) written for the case of absorption of a photon $\hbar\omega$ on the assumption that all states are completely filled in the valence band and free in the conduction band. Since $n_q \to 0$ as the crystal temperature $T \to 0$, Eq. (81) implies that the absorption of a photon $\hbar\omega$ at low temperatures occurs only with the simultaneous emission of a phonon $\mathscr{E}_q$. In considering the generation of negative temperature states by our proposed method, we are concerned primarily with an indirect transition corresponding to recombination accompanied by the simultaneous emission of a photon $\hbar\omega$ and phonon $\mathscr{E}_q$. This transition is the opposite of the transition described by the first term in the expression (81) (with the coefficient B_1).

If we now assume a total inverted population of the valence band and the conduction band, the maximum gain (designated by $\varkappa$) that can be obtained by virtue of induced transitions with simultaneous emission of a photon $\hbar\omega$ and a phonon $\mathscr{E}_q$ is equal to

$$-\varkappa = B_1 (n_q + 1)(\hbar\omega + \mathscr{E}_q - \Delta)^2. \tag{82}$$

This follows from the expression (67) with $f_e(\varepsilon_e) = f_h(\varepsilon_h) = 1$. The condition for obtaining a negative absorption coefficient is in the inequality

$$B_1 (\hbar\omega + \mathscr{E}_q - \Delta)^2 > n\sigma_e + p\sigma_h. \tag{83}$$

The latter inequality is written on the assumption that absorption occurs only at free carriers [it would be more correct on the right side of Eq. (83) to write in directly the experimental absorption coefficient for the created nonequilibrium concentration] and that the temperature is so low that $n_q \ll 1$. The quantity $(\hbar\omega + \mathscr{E}_q - \Delta)^2 = (\Delta E)^2$ is the square of the sum of the energy width near the bottom of the conduction band completely filled with electrons, $f_e(\varepsilon_e) = 1$, and the energy width of the band near the top of the valence band completely filled with holes. Then it follows from the expression (83) that if the quantity $(\Delta E)^2$ increases more rapidly than a linear law as a function of the growth of the nonequilibrium concentration n and p, then it is more suitable for the satisfaction of this inequality to have as large a value of ΔE as possible. All of the above computations are applicable for the absence of degeneracy, while in the case of degeneracy of the nonequilibrium carriers $(\Delta E)^2 \sim$ (n or p)$^{4/3}$, hence it is less appropriate to go to high electron and hole concentrations.

However, it is impossible even in principle to increase these nonequilibrium concentrations without limit, because they can form a negative dielectric constant in the frequency interval with which we are concerned. Moreover, with strong degeneracy, as indicated in Chapter II, there is no need for indirect transitions, and it is better to use recombination without the participation of phonons, which yields a much larger gain. If to all this we add the experimental difficulties associated with the realization of large nonequilibrium concentrations and the required cooling of the sample, it is scarcely feasible to obtain nonequilibrium concentrations for electrons and holes in semiconductors, for example, such as germanium and silicon, higher than 10^{18} cm^{-3}.

For an estimate of the inequality (83), it is necessary to know the coefficient B_1 or to find it from experiment. It could be determined from an observation of the longest-wave absorption edge, when a phonon and photon are absorbed simultaneously, a situation equivalent to the first term on the right side of Eq. (81). However, at low temperatures this term is small, since $n_q \ll 1$. Therefore, at low temperatures, the coefficient B_1 in Eq. (81) is determined experimentally from the dependence of the absorption coefficient on the quantity $(\hbar\omega - \mathscr{E}_q - \Delta)$. If the absorption determined by this relation is calculated in the second approximation of perturbation theory, the coefficient B_2 will differ from the coefficient B_1 only by an insignificant term in the denominator (the phonon energy $\mathscr{E}_q$). For this reason, the empirically obtained coefficient B_2 may be used in place of the coefficient B_1 in (83) for approximate estimations. If now we take into account all that has been argued so far and use the same data as in [39], for germanium, for example ($B_2 \sim 2400$ cm$^{-1} \cdot$ eV^{-2}, $\Delta E = 0.01$ eV, n = $1.9 \cdot 10^{18}$ cm^{-3}, p = $1.0 \cdot 10^{18}$ cm^{-3}), we obtain a gain equal to $\varkappa = 0.24$ cm^{-1}. Assuming the cross section of absorption by free carriers $\sigma_e = 0.3 \cdot 10^{-17}$ cm^{-2} and $\sigma_h = 0.15 \cdot 10^{-17}$ cm^{-2}, we have for the absorption coefficient, according to (76), $\varkappa = 7$ cm^{-1}, i.e., the necessary inequality (83) for the realization of a negative absorption coefficient is not fulfilled at such nonequilibrium carrier concentrations.

Consequently, if the assumed approximate data remain the same at low temperatures (this refers mostly to σ_e and σ_h), then in germanium, clearly, it is impossible to obtain a negative absorption coefficient using indirect interband transitions of current carriers in the free state. The absorption by free carriers could probably be suppressed by placing the semiconductor in a magnetic field. The strength of the magnetic field must be chosen on the basis of the condition of no resonance absorption by the Landau level for the transition frequency with which we are concerned.

3. Negative Absorption Coefficient in Indirect Transitions from Exciton States

Let us investigate the condition for formation of a negative absorption coefficient in the case of indirect transitions from exciton states, when the negative temperature criterion (56) or (62) is met. We observe that the high probability of radiative recombination of an electron and hole bound in an exciton leads us to expect more favorable conditions in this case for the generation of states with a negative absorption coefficient. Moreover, since excitons are classified as Bose particles, more than one exciton may be found in the same state, which is entirely admissible at low temperatures and large excitation densities. On the other hand, there is an added undesirable absorption process, not present with free carriers, namely photoionization of the exciton by electromagnetic radiation. If we treat the exciton as a hydrogen-like atom, the well-known formula for the photoeffect [57] may be used, wherein the electron charge has to be replaced by the effective charge $e/\sqrt{D}$ and the mass of the electron by the reduced mass of the electron and hole. This gives a factor which strongly diminishes the exciton photoionization cross section, reducing it to a negligible quantity. This result is reasonable, insofar as it is well known that the photoeffect cross section falls off sharply with increasing ratio of the quantum energy to the ionization potential. Since the exciton binding energy is very small, amounting approximately to 0.01 eV for semiconductors of the germanium and silicon type, this ratio is roughly equal to 10^2 for quanta with a wavelength $\lambda \sim 1\mu$.

However, an undesirable absorption having a large magnitude is the process of photoionization of the exciton with the simultaneous emission or absorption of a phonon, since the phonon momentum, which may be comparable with the relative momentum of the electron and hole in the final state, causes the matrix element describing this process to be lacking in a rapidly oscillating factor. The phonon energy is enormously larger than the exciton binding energy, hence the photoionization cross section of the exciton with the participation of phonons will be equal to the sum of the cross sections analyzed in the preceding section for absorption by a free electron and hole. As indicated in §1 of Chapter III, at sufficiently low crystal temperatures, according to Eq. (62), the concentration of excitons is greater than the concentration of electrons and holes existing in equilibrium with the excitons. Consequently, it is no longer necessary to account for absorption by free current carriers.

Let the lifetime of the exciton relative to radiation decay be equal to τ in the emission frequency interval from ω to $\omega + \Delta\omega$. Then the number of quanta N radiated by induced emission into unit volume is determined by the equation

$$\left(\frac{dN}{dt}\right)^{+} = \frac{n_r \nu}{\tau},$$

(84)

where ν is the number of excitons per unit volume, n_r is the number of quanta in a definite quantum state.

On the other hand, the number of quanta absorbed per unit volume of the sample is equal to

$$\left(\frac{dN}{dt}\right)^{-} = - c\sigma_i \nu N,$$

(85)

where σ_i is the ionization cross section of the photon. The condition for onset of a negative absorption coefficient leads to the inequality

$$\sigma_i < \frac{\lambda^2}{4D\Delta\omega\tau}.$$

(86)

The lifetime of the exciton τ relative to radiation decay may be related, correct to an order of magnitude, to the probability of recombination of a slow free electron and hole [58].

In fact, the average probability $<W_{ij}>$ introduced in §2 of Chapter III is the reciprocal of the lifetime of an electron and hole in the free state within a volume of 1 cm^3. In the case where they are bound into an exciton, on the other hand, this volume shrinks to $\sim \pi R^3$, where R is the exciton radius. Consequently, the probability of recombination increases such that

$$\frac{1}{\tau} \approx \frac{1}{\pi R^3} \langle W_{ij} \rangle.$$

(87)

The exciton radius R may be assumed approximately equal to $a_0 Dm/m^*$, where $a_0 = \hbar^2/me^2$ is the Bohr radius, D is the dielectric constant, m is the mass of the free electron, and m^* is the reduced effective mass of the electron and hole. Substituting for silicon D ~ 12, $m/m^* \sim 2.5$, and assuming at room temperature, in accordance with [53], that $<W_{ij}> \sim 2 \cdot 10^{-15}$ cm$^3 \cdot$ sec^{-1}, we obtain $\tau^{-1} \sim 2 \cdot 10^5$ sec^{-1}. The experimental value of the exciton lifetime, determined in [45], comprises a value $\tau \sim 6 \cdot 10^{-5}$ sec, which agrees satisfactorily with Eq. (87).

The condition (86) suggests the favorability of having as small a value as possible for the product $\Delta\omega\tau$, i.e., as small a lifetime as possible with respect to radiative transition and as narrow an energy spectrum as possible on the part of the radiation. The width of the exciton emission line is determined down to low temperatures by the kinetic energy of the excitons and is of the order kT. However, there are indications [59] that the width of the exciton line cannot be made as narrow as one wishes, but has a limit, below which the line width is independent of the temperature.

A numerical estimate based on Eq. (86) for silicon ($\Delta\omega \sim 10^{12}$ sec^{-1} for T $\sim$ 10°K) yields

$$\sigma_i < 0.4\cdot10^{-17}\,\text{cm}^2. \tag{88}$$

Consequently, the sum of the cross sections for absorption of a photon $\hbar\omega$ by a free electron and hole in silicon must be smaller than $0.4 \cdot 10^{-17}$ cm^2, which, according to available experimental data [50], is true. This means that the condition for a negative absorption coefficient in transitions from exciton states is realized in this case. As in the case of free carriers, a magnetic field can be used to diminish the cross section of photon absorption by excitons at the relevant frequencies.

We have examined the low-temperature case, when the concentration of excitons is greater than the free carrier concentration. A detailed comparison of the probabilities of indirect radiative recombination from the exciton state and from the free carrier state in diverse cases is offered in [55, 60].

It is also possible to assess the role of other processes of absorption of radiation by impurities, defects of the crystal lattice, and by the lattice itself. Thus, the number of quanta absorbed per unit volume in unit time is equal to

$$\left(\frac{dN}{dt}\right)^- = -c\varkappa_c N, \tag{89}$$

where $\varkappa_c$ is the total absorption coefficient at the wavelength corresponding to indirect transition. For the realization of a negative absorption coefficient, we determine the condition for the exciton concentration from (84) and (89):

$$\nu > 4D\ \Delta\omega\tau\lambda^{-2}\varkappa_c. \tag{90}$$

A numerical estimate by means of the values of the parameters assumed in (88) yields

$$\nu > 2.4\cdot10^{17}\ \varkappa_c. \tag{91}$$

We note that the condition for realization of a negative absorption coefficient in the case of absorption by an exciton depends only on the relation of the induced recombination process and exciton photoionization, but not on the exciton concentration; other absorption processes, by contrast, as indicated by the expression (91), can be masked by an increase in the exciton concentration. Consequently, in the case of indirect transitions, fairly small nonequilibrium concentrations will result in negative temperature states, but the absorption for intraband transitions does not provide a negative absorption coefficient at such low nonequilibrium concentrations. However, the processes associated with intraband absorption have practically no effect on the conductivity, because they do not alter the number of free carriers. In the negative temperature state, on the other hand, induced transitions reduce the number of free carriers and lead to a decrease in the conductivity (negative photoconductivity).

Experiments were conducted in [15] on the creation and observation of negative temperature states in silicon. A sample at a temperature of 4°K was irradiated with intense light at a wavelength of less than 0.7 μ, which increased the conductivity of the sample appreciably. Upon additional irradiation with weak monochromatic light, a reduction in the conductivity was observed for a number of the samples in a narrow interval of wavelengths in the vicinity of 1.1 μ.

CHAPTER IV

REALIZATION OF NEGATIVE TEMPERATURE STATES
IN SEMICONDUCTORS FOR OPTICAL AND ELECTRONIC
EXCITATION SOURCES

In the first chapter, we analyzed in detail the method of obtaining negative temperature states with the cumulative ionization of a homogeneous and an impurity semiconductor. As noted previously, the fundamental drawbacks of the method are: the ability to function only in the pulsed mode, high electric field strengths, and large energy losses.

In the second chapter, we discussed the method of injection of nonequilibrium carriers through a p−n junction of degenerate semiconductors. This method is devoid of the energy losses just mentioned and is capable of giving an efficiency near unity. However, the region of radiation generation still has such small dimensions that it is difficult to produce generated radiation with a high power using this method of excitation.

In the present chapter, we investigate the methods of creating negative temperature states in semiconductors by means of optical and electronic excitation. Although, as will be apparent presently, it is impracticable with such methods of excitation to obtain such a high efficiency as in the method of injection through a p−n junction, it is still possible to realize a negative temperature in a large region of the sample and hence (particularly in the case of electronic excitation) to produce a high generated power.

1.   Method of Optical Excitation of Semiconductors

One of the first methods proposed for the generation of negative temperature states in semiconductors was the method of optical excitation. At the First and Second International Conferences on Quantum Electronics, Prof. B. Lax [6, 7] suggested obtaining negative temperature by the selective excitation of carriers at the cyclotron resonance level using light from a powerful lamp monochromatized by a light-force spectrometer. Moreover, he showed that the small lifetime of the carriers at the Landau levels ($\sim 10^{-12}$ sec) require large powers for excitation.

The number of quanta i in $cm^{-2} \cdot sec^{-1}$ required for the formation of n pairs $\cdot cm^{-3}$ may be ascertained as follows: $i \sim n/\tau_c \varkappa$, where $\varkappa$ is the absorption coefficient for incident photons and τ_c is the lifetime of nonequilibrium pairs. It would be more correct to adopt, rather than $\varkappa$, the reciprocal diffusion length L_D^{-1}, on which the excitation region is actually formed. We investigated the most favorable case, when an electron−hole pair is formed at each quantum; of course, the energy of the incident quanta must be greater than some minimum energy for the generation of an electron and hole.

Consequently, if a negative temperature is created between levels yielding a direct transition, then, as implied by §1 of Chapter I, the density n must be such that the separation of the Fermi quasi-levels exceeds the width of the forbidden band (or the distance from the impurity level to the band from which the relevant transition originates). It is clear that in this case n is large (of course, for not too low temperatures), and obviously conventional light sources can only be used with a highly efficient utilization of their radiation for excitation and then only in a pulsed mode (for example, for n $\sim 10^{17}$ cm^{-3}, $\tau_c \sim 10^{-6}$ sec, and L$_D \sim 10^{-4}$ cm, we have i $\sim 10^{-19}$ quanta $\cdot$ cm$^{-2} \cdot$ sec^{-1}).

The realization of negative temperature states by the method of optical pumping is facilitated if the indirect transitions investigated in the preceding chapter are used, where there is no need to have carrier degeneracy in the bands. As already indicated in §1 of Chapter III, an intensity i $\sim 10^{13}$ quanta $\cdot$ cm$^{-2} \cdot$ sec^{-1} is sufficient for the formation of negative temperature in indirect transitions. For the realization of a negative absorption coefficient, however, according to §2 of Chapter III, large concentrations and, hence, large values of i, are required. For indirect transitions from exciton states, provided the conditions (86) and (90) are fulfilled, the nonequilibrium concentration of excitons and, consequently, the intensity i may be small, so that it becomes possible to operate in a continuous mode.

As in the case of pulsed excitation of homogeneous semiconductors by an electric field, the formation of a negative temperature state with optical excitation is possible only if the photon-generated electrons and holes are slowed down as the result of interaction with the crystal lattice, transferring to lower energy levels, i.e., if the slowing-down time is less than the lifetime. Consequently, all of the energy of the electrons and holes initially generated by the photon, this energy exceeding the energy of the photon emitted in recombination, is released in the form of heat and produces undesirable heating of the crystal lattice.

A disadvantage of the optical method of excitation by means of conventional light sources is the fact that they yield a high power and broad energy spectrum, so that short-wave photons result in high energy losses upon relaxation to the lower levels. Nonmonochromatic light sources are not focused on small areas, thus impeding the generation of large excitation densities. Moreover, proper illumination of the surface of the excited sample is difficult with a broad spectrum, which means that considerable reflection takes place on the part of the incident radiation.

The use of radiation from lasers (for example, a ruby laser) in order to obtain negative temperature states in semiconductors eliminates a large part of the difficulties mentioned above in connection with the application of ordinary light sources. As a matter of fact, the radiation from a laser is coherent and monochromatic, hence it can be focused and yields an exceedingly high excitation density, and the attendant illumination of the sample surface almost totally eliminates the reflection effect. It is possible, by suitable choice of radiation frequency for the laser and width of the forbidden band of the semiconductor, to attain minimum energy losses associated with relaxation of the generated nonequilibrium carriers, these losses being equal to the difference in energy of the absorbed quantum and the energy of the emitted quantum for each incident quantum.

The usual shortcoming of the optical method of exciting semiconductors is the fact that, due to the large absorption coefficient with respect to the incident light, only a thin layer is excited near the surface (about 10^{-4} cm), where various processes of nonradiative recombination are particularly strong, diminishing the lifetime τ_c. However, if the incident light is absorbed by means of indirect transitions, a layer of considerable thickness can be excited, and the high radiation intensity of a laser promotes the necessary degree of excitation. For example, a very suitable source is the radiation from a glass laser with a neodymium impurity [61] (wavelength about 1.06 μ) for the excitation of silicon. The recombination radiation of semiconductors at various temperatures under the influence of radiation from a ruby laser (wavelength 6943 Å) was first observed in [62].

Especially promising is the application of lasers with a modulated quality factor [63, 64], which make it possible to create very high excitation densities over a short period of time (10^{-7} to 10^{-8} sec). If the duration of the pulse from the driving oscillator is shorter than the lifetime of the generated nonequilibrium carriers, the number of pairs formed is equal to

$$n \approx \frac{P\Delta t}{SL_D\chi},$$
(92)

where P is the power of the laser, S is the area of the focused beam, Δt is the pulse duration of the laser, and χ is the energy expended in formation of the pair.

If in (92) we let [65] $P \sim 3 \cdot 10^7$ W for $\Delta t \sim 0.3 \cdot 10^{-7}$ sec, $S \sim 10^{-1}$ cm^2, $L_D \sim 10^{-4}$ cm, $\varkappa \sim 1$ eV, we obtain $n \sim 10^{19}$ cm^{-3}, i.e., in almost all semiconductors the lifetime will be short. In the event that the lifetime $\tau_C < \Delta t$, Δt must be replaced by τ_C in Eq. (92).

We note that the use of an optical laser for the creation of a semiconductor laser results in the transformation of shorter-wavelength radiation into longer-wave radiation with inevitable energy losses. However, the use of radiation from lasers for the excitation of semiconductors could prove very useful for the selection of suitable semiconductor materials to be used later in lasers driven by other methods.

2. Excitation of Semiconductors by an Electron Beam

The application of an electron beam for the excitation of semiconductors, as opposed to the optical pumping method, enables one to produce negative temperature states for semiconductors with various widths of the forbidden band, i.e., to produce coherent generation in the range from the far-infrared to the ultraviolet region. Electron beams are capable of supplying an energy sufficient for the excitation of a considerable sample thickness, the excitation intensity can be made very high, and the device can be made to function in the short-pulse mode, thus eliminating any substantial heating of the crystal lattice.

The energy of the primary electron beam must be chosen so as to stimulate a layer of sufficient thickness, but without resulting in the formation of unwanted effects, which raise the probability of nonradiative transitions and tend to impair the optical homogeneity of the semiconductor. As implied by [66], the process of defect formation in such semiconductors as germanium and silicon becomes appreciable at electron energies of about 0.5 MeV. It is inappropriate, therefore, to use electron beams with an energy higher than 0.5 MeV. For electrons with energy up to 3 MeV, we have a fairly precise equation [67] linking the penetration depth R (in g per cm^{-2}) with the energy E_0 (in MeV):

$$R = 0.11 \left(\sqrt{1 + 22.4E_0^2} - 1 \right),$$

i.e., in germanium and silicon, the electrons penetrate to a depth of ~ 1 mm when their energy is on the order of 0.5 MeV.

For excitation with an electron beam, losses will be unavoidable, even if a photon is emitted in the recombination of each pair formed. The possibility of these losses was indicated by the author in [68] and was subsequently analyzed [69] to the extent that the maximum energy efficiency of a semiconductor quantum oscillator driven by an electron beam could be estimated. Inside a crystal, fast electrons form electrons and holes, the kinetic energy of which is many times the thermal energy corresponding to the lattice temperature. This follows from the fact that the minimum kinetic energy required for the formation of an electron (or hole) of one pair (electron and hole), as will be shown presently, is not equal to the width of the forbidden band, and a sizable portion of the energy must transform to kinetic energy of the particles formed. In this case, the particle itself and each particle from the generated pair can have an energy insufficient for the formation of a new pair, but still many times larger than the thermal energy. Such particles enter into equilibrium with the lattice (thermolysis) prior to their recombination, imparting to the lattice an excess energy, over and above the thermal energy, if the slowing-down time is less than the lifetime (it is shown in §1 of Chapter V that this relation is satisfied).

In this case, even if all of the generated electrons and holes recombine with emission, the energy efficiency will be less than unity.

It will be demonstrated below that the energy losses due solely to the mechanism indicated above can reach 75%, the most probable losses amounting to 60%. It is important to emphasize that these losses do not depend on the purity of the semiconductor, the energy of the primary beam or, clearly, on the particular semiconductor material used. It is logical to assume, therefore, that such energy losses in electronic excitation are theoretically unavoidable.

The problem of the complete exchange of the energy of the primary electrons for energy above 1 keV inside a solid is very complex in view of the several stages involved in this process (via secondary, tertiary, and higher-generation electrons) and is certainly still far from solution. Without going into this process, we hypothesize that as long as the energy of all electrons and holes (primary, secondary, etc.) in a solid is adequate for the ejection of an electron from the valence band into the conduction band, ionization processes are the main ones involved. This is justifiable in that the principal energy losses of electrons in passage through a solid in the energy region we are concerned with (up to 100 keV) are associated with ionization [42]. Consequently, we assume that the action of the primary electron beam in the conduction band (in the valence band for holes) of the crystal is such as to form electrons instantaneously† with a kinetic energy less than the energy necessary for ionization.

In [26] the probability of ionization w_i by an electron with energy ε_0 is evaluated in the case of the Born approximation

$$w_i \approx \frac{e^4 m}{\hbar^3} \left(\frac{\varepsilon_0 - \varepsilon_i}{\varepsilon_i} \right)^2,$$

where ε_i is the threshold ionization energy. Consequently, the time in which ionization takes place ($\varepsilon_0 \sim 2\varepsilon_i$) is equal in order of magnitude to $\tau_i \sim 10^{-15}$ sec which, of course, is much shorter than all the other characteristic times.

All fast electrons (and holes) whose kinetic energy is less than the quantity defined above (ionization threshold) cannot form an electron−hole pair. Let us see how large this energy actually is. Assume that we have a fast electron (hole), which has a quasi-momentum $\mathbf{p}_0$ in the conduction (valence) band and a kinetic energy, read from the bottom,

$$\varepsilon(\mathbf{p}_0) = \frac{\mathbf{p}_0^2}{2m_e},$$

where m_e is the effective mass of the electron. After ionization, two additional particles are formed, an electron with momentum $\mathbf{p}_{e2}$ and a hole with momentum $\mathbf{p}_h$; the momentum of the primary electron after ionization is designated $\mathbf{p}_{e1}$.

It follows from the conservation of energy and momentum that

$$\mathbf{p}_0 = \mathbf{p}_{e1} + \mathbf{p}_{e2} + \mathbf{p}_h, \tag{93}$$

$$\varepsilon(\mathbf{p}_0) = \varepsilon(\mathbf{p}_{e1}) + \varepsilon(\mathbf{p}_{e2}) + \varepsilon(\mathbf{p}_h) + \Delta, \tag{94}$$

where Δ is the width of the forbidden band.

The ionization threshold is determined by the condition

$$\varepsilon_{ie} \equiv \varepsilon_{\min}(\mathbf{p}_0) \equiv \min\left[\varepsilon(\mathbf{p}_{e1}) + \varepsilon(\mathbf{p}_{e2}) + \varepsilon(\mathbf{p}_h) + \Delta\right]. \tag{95}$$

----

† The characteristic time of the ionization losses is less than all the other characteristic times that we have considered.

The minimum condition on the right side of the equation means that $\nabla\varepsilon(\mathbf{p}_{e1}) = \nabla\varepsilon(\mathbf{p}_{e2}) = \nabla\varepsilon(\mathbf{p}_h) = \mathbf{V}$, at the ionization threshold, i.e., the velocities of all the final particles are the same. Next, making use of (93) and (94), we obtain

$$\varepsilon_{ie} = \Delta\left(1 + \frac{m_e}{m_e + m_h}\right). \tag{96}$$

If the effective masses of the electron and hole are equal, it follows from (96) that $\varepsilon_{ie} = \frac{3}{2}\Delta$, i.e., the ionization threshold is one and a half times the width of the forbidden band. Electrons from the conduction band with a kinetic energy less than ε_i do not participate in the multiplication process.† These electrons can collide with the lattice of the semiconductor, slowing down to thermal energies, or they can recombine radiatively and nonradiatively. It will be shown in §1 of Chapter V that the former process occurs far more rapidly than the latter; it may be assumed without appreciable error, therefore, that the processes ostensibly proceed in series; the first manages to go to completion before the second has a chance to attain any magnitude. The kinetic energy acquired in ionization by a superthermal electron (hole) will be imparted to the lattice. This energy is dissipated in the form of heat and comprises a portion of the unavoidable energy losses incurred in the excitation of a semiconductor by an electron beam.

In order to calculate quantitatively the fraction of these losses, it is necessary to know the electron (hole) energy spectrum acquired as the result of ionization. This spectrum, as mentioned earlier, occupies the interval from the bottom of the conduction band $\varepsilon = 0$ to the ionization threshold $\varepsilon = \varepsilon_i$. The exact energy distribution of the electrons can only be obtained by solving the problem of the complete exchange of the energy of the primary, secondary, and higher-generation electrons (holes).

If all the electrons and holes formed as the result of ionization had an energy near the ionization threshold (ε_{ie} for electrons, ε_{ih} for holes), the energy efficiency η would be minimized. In this case,

$$\eta_{\min} = \frac{\Delta}{\Delta + \varepsilon_{ie} + \varepsilon_{ih}}. \tag{97}$$

If we take the values of ε_{ie} and ε_{ih} from (96) for Eq. (97), we obtain $\eta_{\min} = 0.25$.

Let $f_e(\varepsilon)\varepsilon^{1/2}d\varepsilon$ be the number of electrons from the conduction band with energy in the interval‡ from ε to $\varepsilon + d\varepsilon$, so that

$$\int_0^{\varepsilon_{ie}} f_e(\varepsilon)\,\varepsilon^{1/2}\,d\varepsilon = N_0, \tag{98}$$

where N_0 is the total number of electrons and holes formed in the conduction band as the result of ionization. An analogous relation applies to holes. The energy efficiency of cathode luminescence is equal to

$$\eta = \frac{\Delta N_0}{\Delta N_0 + \int_0^{\varepsilon_{ie}} \varepsilon f_e(\varepsilon)\,\varepsilon^{1/2}\,d\varepsilon + \int_0^{\varepsilon_{ih}} \varepsilon f_h(\varepsilon)\,\varepsilon^{1/2}\,d\varepsilon}. \tag{99}$$

---

† Actually, electrons in the energy interval from Δ to ε_i can form a pair only with the simultaneous emission of a phonon. The role of such processes will be evaluated later.

‡ We assume that the relation $\varepsilon = p^2/2m$ exists between the energy and momentum, i.e., that the effective mass approximation is valid; in this case, the density of the levels is proportional to $\sqrt{\varepsilon}$.

As mentioned above, the energy distribution of the electrons and holes formed as the result of ionization is not known, and there does not exist at the present time any quantitative calculation of this distribution. However, it may be stated that the distribution for which $\eta = 1$, i.e., such that all electrons are formed at the bottom of the conduction band (and holes, accordingly, at the top of the valence band), is unrealizable, since the probability of such a process is zero (it is proportional to the density of final states, which goes to zero at the edges of the bands). Moreover, ionization is inevitably accompanied by the transfer of a sizable portion of the energy of the ionizing particle into kinetic energy of the generated particles. The magnitude of the energy efficiency, according to (99), may be determined if the nature of the distribution $f(\varepsilon)$ of electrons and holes obtained as the result of cumulative ionization is postulated.

We postulate that the incidence of electrons in the levels of the conduction band from the bottom of the band to an energy ε_{ie} is equally probable, i.e.,

$$f_e(\varepsilon_e) = C_e \quad \text{for} \quad 0 \leqslant \varepsilon_e \leqslant \varepsilon_{ie};$$

$$f_h(\varepsilon_h) = C_h \quad \text{for} \quad 0 \leqslant \varepsilon_h \leqslant \varepsilon_{ih}.$$

This assumption, clearly, is close to reality, since the energies of the exciting primary electrons $\varepsilon \gg \varepsilon_e(\varepsilon_h)$ until the last stages of ionization. Even if the true distribution function is constant, this does not yield a significantly different numerical value for η, due to the integral dependence of η on the function f. The existing experimental data on secondary emission [70-71] and on the direct transmission of a beam evinces the formation of electrons in a solid with energies hundreds of times the thermal energy (electrons with energy from 1 to 10 eV), which is consistent with our postulate.

According to the condition (98), we have

$$\left.\begin{array}{l} C_e = {}^3\!/_2 N_0 (\varepsilon_{ic})^{-3\!/_2}, \\ C_h = {}^3\!/_2 N_0 (\varepsilon_{ih})^{-3\!/_2}. \end{array}\right\} \tag{100}$$

After the given postulate, the energy efficiency in the case of losses solely due to thermolysis is determined as follows:

$$\eta = \frac{\Delta N_0}{\Delta N_0 + C_e \int\limits_0^{\varepsilon_{ic}} \varepsilon^{3\!/_2} d\varepsilon + C_h \int\limits_0^{\varepsilon_{ih}} \varepsilon^{3\!/_2} d\varepsilon}.$$

From this, after some straightforward calculations, we obtain

$$\eta = \frac{1}{1 + {}^3\!/_5 \left(\dfrac{\varepsilon_{ie} + \varepsilon_{ih}}{\Delta} \right)}. \tag{101}$$

Substituting the expressions for ε_{ie} and ε_{ih} obtained in (96) into (101), we find a value of $\eta = {}^5\!/_{14}$. Consequently, the energy efficiency for excitation by an electron beam cannot be higher than 36% within the framework of the given postulates. As apparent from the arguments put forth above, η depends for the most part only on ε_{ie} and ε_{ih} but not on the energy of the primary beam, a situation that stands in good agreement with the experimental data of [72, 73]. The mean energy expended in the formation of a pair, in our case equal to about 3Δ, also agrees fairly well with the experimental data of [73]. In the calculation of η according to Eq. (101), we inserted the values of ε_{ie} and ε_{ih} obtained on the assumption of direct creation of an electron and hole of an additional pair. As noted previously, the law of conservation of momentum gives values for ε_{ie} and ε_{ih} greater than the width of the forbidden band Δ. However, if in the creation of the pair a phonon is simultaneously emitted or absorbed, the quantities ε_{ie} and ε_{ih} will be equal to Δ. In this case, the energy efficiency attains a value $\eta = \frac{5}{11}$.

The role of creation processes with the participation of phonons has been subjected to a detailed analysis in [74]. However, from our point of view, Shockley's calculation of the mean energy required for the formation of a pair of high-energy particles is not convincing. Shockley postulated that every ionization process is accompanied by the emission of a phonon; inasmuch as this process is one of low probability, the emission of a large number of phonons occurs (17 in silicon, 57 in germanium) before ionization actually takes place. In his calculations, therefore, the energy losses due to photon emission in multiplication comprise about half the total losses. It seems strange to us that it should be necessary, for a fast particle, which has a high probability [26] ($\tau_i \sim 10^{-15}$ sec) of forming a pair without the participation of a phonon, to take into account the emission of phonons (the phonon emission time cannot be less than 10^{-12} sec), and even more so the process of ionization with the emission of phonon which, according to Shockley's results, is $\frac{1}{17}$ to $\frac{1}{50}$ as probable as the process of phonon emission.

The foregoing arguments indicate that as long as all the electrons and holes formed in a solid have an energy sufficient for direct ionization (without a phonon participating), the particle energy will be allocated solely for the creation of pairs. Further energy losses will occur due to the emission of phonons in thermolysis of the electrons and holes. In principle, if the quantities ε_{ie} and ε_{ih} are greater than Δ, it should be required to take into account the process of ionization with emission of a phonon for electrons and holes having a kinetic energy in the interval from Δ to ε_{ie}, ε_{ih}. However, as Shockley's results indicate, the electrons and holes are able to escape from this interval into an interval of energies less than Δ before ionization takes place with the emission of a phonon. Consequently, we arrive at the conclusion that ionization processes involving the simultaneous emission of phonons do not play a significant role for the energy losses of fast particles in a semiconductor. Moreover, in our calculations, we obtain agreement with experiment for the amount of energy consumed in the formation of a pair without the incorporation of any constants.

As will be shown in the next chapter, at the carrier concentrations required for the formation of negative temperature states, the electrons and holes produced as the result of multiple ionization exchange energy due to mutual collisions within very short periods (shorter than the time for collision with the lattice) and are distributed according to Fermi statistics with some effective temperature Θ different from the lattice temperature T. Inasmuch as we have shown that at least half the energy admitted by the electron beam is imparted to the crystal lattice by the generated electrons and holes, the temperature Θ must differ from the temperature T to such an extent that the power delivered from the electrons and holes to the lattice in the quasistationary regime would be equal to the power released in recombination.

For interaction exclusively with acoustic vibrations of the lattice, i.e., when $\Theta < \Theta_0$, the power transmitted by the electron gas to the lattice is determined by Eq. (152). For sufficiently strong degeneracy, when $\mu/k\Theta \gg 1$, we obtain from Eq. (152) the energy transmitted to the lattice per unit time:

$$\left(\frac{dE}{dt}\right)_{\text{lat}} = \left(\frac{3}{\pi}\right)^{4/3} \frac{\pi^2}{2\sqrt{2}} \frac{a_0 \hbar n^{4/3}}{\sqrt{m}} k(\Theta - T),$$

where the same notation is used for the physical variables as in Eq. (152). On the other hand, the power released in recombination of n electrons $\cdot$ cm^{-3} is equal to

$$\left(\frac{dE}{dt}\right)_{\text{rec}} \approx \frac{n\Delta}{\tau_c},$$

where τ_c is the lifetime in the recombination of an electron with a hole and Δ is the energy width of the forbidden band. Equating these powers, we obtain an expression for the difference between the temperatures of the electron gas and semiconductor lattice:

$$k\,(\Theta - T) \approx \frac{2\sqrt{2}}{\pi^2}\left(\frac{\pi}{3}\right)^{4/3}\frac{\sqrt{m}}{a_0\hbar n^{1/3}}\frac{\Delta}{\tau_c}\,.$$

As implied by the expressions (136) and (137) for the constant a_0, its value is inversely proportional to the carrier mobility, so that, consequently, the temperature of carriers with a high mobility will be low. If we use the value of a_0 for germanium and assume a lifetime $\tau_c \sim 10^{-6}$ sec and $n \sim 10^{18}$ cm^{-3}, the temperature of the electrons will only differ by a few degrees from the temperature of the crystal lattice. However, if the lifetime is considerably smaller than 10^{-6} sec, as for example in gallium arsenide, the temperature of the electrons may increase so much that the degeneracy of the current carriers is liquidated and, consequently, it becomes impossible to obtain a negative temperature state in the quasi-stationary regime. In this case, excitation can be realized only by means of short pulses (shorter than the lifetime) or degeneracy is obtained with very sudden removal of the driving electron beam (front shorter than the lifetime).

We point out that the temperature Θ increases rather quickly with the admitted power until it reaches a value of about Θ_0 (Debye temperature). With a further increase in temperature, most of the power will be dissipated in optical rather than acoustic vibrations of the lattice. As seen in Chapter V, a more vigorous transfer of energy to the lattice takes place, and the temperature Θ differs very little from the temperature Θ_0, even with a large excitation power. However, it would seem difficult to us to realize conditions such that the nonequilibrium carrier concentrations required for degeneracy could be created for a temperature of the electron gas $\Theta \sim \Theta_0$.

The above result for the heating of the electron gas in excitation by an electron beam is also valid in the case of powerful optical pumping, for example, when using the radiation from lasers, the only difference being that the quantity Δ has to be replaced by $(\hbar\omega - \Delta)$, where $\hbar\omega$ is the energy of the exciting quanta. In this case, it is not necessary to allow for heating due to absorption of the exciting radiation by free carriers, because this effect is weakened by comparison with the heating associated with interband absorption in the same ratio as the coefficients of absorption by free carriers to the absorption coefficient for interband transitions.

CHAPTER V

SLOWING–DOWN TIME OF NONEQUILIBRIUM CURRENT CARRIERS IN SEMICONDUCTORS

As remarked in the preceding chapters, the determination of the slowing-down time of nonequilibrium current carriers in semiconductors is of paramount importance in creating negative temperature states in semiconductors.

In a number of the methods discussed in Chapters I, III, and IV, the realization of negative temperature states was based on the notion of a small slowing-down time in comparison with the electron–hole recombination time. The slowing-down time is understood to mean the time required for the carrier to lose the excess amount of kinetic energy it has over and above thermal energy by interaction with the crystal lattice. Rapid deceleration of the nonequilibrium current carriers fosters a high degree of occupation of the energy levels near the edges of the bands by a reduction in the temperature of the crystal.

In sufficiently pure and perfect crystals, the retardation of the current carriers is determined by scattering by the lattice vibrations (optical and acoustic). The role of impurities turns out not to be too significant, as the carrier, when scattered by impurity atoms, transmits to the atom only a small part of its energy, on the order of m/M, where m and M are the masses of the carrier and impurity atom, respectively. In the present chapter, therefore, we will examine the slowing down of current carriers only in connection with interaction with the lattice of a pure homogeneous semiconductor.

1. Derivation of the Fundamental Kinetic Equation

We begin with a derivation of the kinetic equation for the distribution function of electrons† in the crystal, taking into account the Fermi degeneracy of the electron gas, since it becomes significant in the final stages of retardation. In the kinetic equation, we omit the term describing collisions of the electrons with one another, as such a process does not lead directly to any pertinent reduction in the mean energy of the electrons, but only gives a contribution to the slowing-down process indirectly, through the form of the distribution function, which will be accounted for later.

The kinetic equation for the electron distribution function $f(\mathbf{p})$, taking into account interaction with the lattice vibrations [75], has the form

$$\frac{\partial f(\mathbf{p})}{\partial t} = \sum_{\mathbf{q}} \{-\rho_a(\mathbf{q}) f(\mathbf{p}) [1 - f(\mathbf{p} + \mathbf{q})] + \rho_e(\mathbf{q}) f(\mathbf{p} + \mathbf{q}) [1 - f(\mathbf{p})] -$$

$$- \rho_e(\mathbf{q}) f(\mathbf{p}) [1 - f(\mathbf{p} - \mathbf{q})] + \rho_a(\mathbf{q}) f(\mathbf{p} - \mathbf{q}) [1 - f(\mathbf{p})]\}, \tag{102}$$

where $\rho_a(\mathbf{q})$ is the probability of a phonon with momentum $\mathbf{q}$ being absorbed by an electron and $\rho_e(\mathbf{q})$ is the probability of a phonon with momentum $\mathbf{q}$ being emitted. The summation on the right side of Eq. (102) is taken over all possible values of the momentum $\mathbf{q}$, both positive and negative.

The probabilities of absorption and emission of a phonon, ρ_a and ρ_e, are equal to

$$\rho_a = \frac{2\pi}{\hbar} B(q) n_q \delta(\varepsilon_2 - \varepsilon_1 - \hbar\omega_q), \tag{103}$$

$$\rho_e = \frac{2\pi}{\hbar} B(q) (n_q + 1) \delta(\varepsilon_2 - \varepsilon_1 + \hbar\omega_q), \tag{104}$$

where B(q) is the square of the matrix element for interaction of an electron with a phonon, n_q is the number of phonons in the crystal with momentum $\mathbf{q}$ and energy $\hbar\omega_q$, and ε_2 and ε_1 are the final and initial energies of the electron.

We go from summation in Eq. (102) to integration over the phonon momentum $\mathbf{q}$. For this we need to replace the symbol $\sum_{\mathbf{q}}$ by

$$\int \frac{V}{(2\pi\hbar)^3} q^2 dq\, d(\cos\theta)\, d\varphi, \tag{105}$$

where V is the volume of the crystal, θ is the polar angle, and φ is the azimuthal angle. We will consider for simplicity that the dispersion law for the current carriers in a semiconductor is spherically symmetrical with a square-law dependence of the energy on the quasi-momentum, i.e., it has the form $\varepsilon(\mathbf{p}) = p^2/2m$, where m is the effective mass of the carrier. With this

---

† In the ensuing paragraphs, we discuss only electrons in the conduction band, but all the results are entirely valid for the holes in the valence band as well.

assumption in mind, substituting (103)-(105) into (102), we carry out the integration over the angle θ, while integration over φ merely yields a factor 2π, due to cylindrical symmetry. We first consider in the expression (102) the sum of the first and second terms, which have the same argument of the delta function. From the conservation of energy and momentum we have

$$\varepsilon(\mathbf{p}+\mathbf{q}) = \varepsilon(\mathbf{p}) + \hbar\omega_q, \tag{106}$$

i.e.,

$$\frac{(\mathbf{p}+\mathbf{q})^2}{2m} - \frac{\mathbf{p}^2}{2m} = \hbar\omega_q. \tag{107}$$

The latter equation may be rewritten in the form

$$\frac{pq\cos\theta}{m} + \frac{q^2}{2m} = \hbar\omega_q, \tag{108}$$

where θ is the angle between the vectors $\mathbf{p}$ and $\mathbf{q}$, i.e., the polar angle of our chosen coordinate system. Now integration over the angle θ is readily performed:

$$\int_{-1}^{1} \delta[\varepsilon(\mathbf{p}+\mathbf{q}) - \varepsilon(\mathbf{p}) - \hbar\omega_q] \, d(\cos\theta) = \int_{-1}^{1} \delta\left(\frac{pq\cos\theta}{m} - \frac{q^2}{2m} - \hbar\omega_q\right) d(\cos\theta) = \frac{m}{pq}. \tag{109}$$

Bearing this result in mind, the first two terms of Eq. (102) may be rewritten as follows:

$$\sum_q \{-\rho_a(\mathbf{q})f(\mathbf{p})[1 - f(\mathbf{p}+\mathbf{q})] + \rho_e(\mathbf{q})[1 - f(\mathbf{p})]f(\mathbf{p}+\mathbf{q})\} =$$
$$= \frac{Vm}{2\pi\hbar^4 p} \int B(q)\, q\, dq\, \{-n_q f(\mathbf{p})[1 - f(\mathbf{p}+\mathbf{q})] + (n_q+1)f(\mathbf{p}+\mathbf{q})[1 - f(\mathbf{p})]\}. \tag{110}$$

The limits of integration over the absolute value of the phonon momentum q must be chosen from the condition $-1 \leq \cos\theta \leq 1$, determined according to the equation (108). The value of the upper limit is found with $\cos\theta = -1$ (assuming that $\hbar\omega_q$ grows more slowly than q^2), i.e., the maximum momentum q_2 is determined from the condition

$$q_2 - \frac{2m\hbar\omega_q}{q_2} = 2p. \tag{111}$$

This equation includes the unknown function $\hbar\omega_q$, hence it cannot be solved in general form. However, for the cases of practical interest involving acoustic and optical vibrations of the lattice, $\hbar\omega_q = \hbar uq$, where u is the velocity of sound, or $\hbar\omega_q = \text{const}$, so that the value of q_2 may be found exactly:

$$\left.\begin{array}{l} q_2^{ac} = 2p + 2mu, \\ q_2^{op} = p + \sqrt{p^2 + 2m\hbar\omega_0}, \end{array}\right\} \tag{112}$$

where $\hbar\omega_q$ is the energy of the optical phonon.

If in this equation $q_2/\hbar$ exceeds the reciprocal lattice vector K, the upper limit of integration is equal to $\hbar K$. The lower limit of integration may be found from the condition $\cos\theta = 1$, provided only that the value $q_1 = 0$ does not meet the condition (108) for values of $\cos\theta < 1$. In such event, the lower limit is zero; this is true, in particular, for acoustic vibrations of the lattice, where the limiting value of $\cos\theta$ is equal to $mu/p = u/v$, v is the velocity of the electron (we are not considering "cold" electrons, when $u \approx v$). Consequently, for the lower limit in the case of scattering by acoustic vibrations we can write

$$q_1^{\text{ac}} = 0, \left.\begin{array}{c}\\ \\\end{array}\right\}$$
$$q_1^{\text{op}} = -p + \sqrt{p^2 + 2m\hbar\omega_0.}$$

(113)

We now investigate the last two terms of Eq. (102). The conservation of energy gives

$$\varepsilon(\mathbf{p}) - \varepsilon(\mathbf{p} - \mathbf{q}) = \hbar\omega_{\mathbf{q}}.$$

(114)

As in the preceding case, we introduce the angle between the vectors $\mathbf{p}$ and $\mathbf{q}$:

$$\frac{pq\cos\theta}{m} - \frac{q^2}{2m} = \hbar\omega_q.$$

(115)

Integrating over θ in Eq. (115), we obtain

$$\int_{-1}^{1} \delta[\varepsilon(\mathbf{p}) - \varepsilon(\mathbf{p} - \mathbf{q}) - \hbar\omega_q]\, d(\cos\theta) = \frac{m}{pq}.$$

(116)

Now the last two terms of Eq. (102) may be written as

$$\sum_{\mathbf{q}} \{-\rho_e(\mathbf{q})\, f(\mathbf{p})[1 - f(\mathbf{p} - \mathbf{q})] + \rho_a(\mathbf{q})\, f(p-q)[1 - f(\mathbf{p})]\} =$$
$$= \frac{Vm}{2\pi\hbar^4 p} \int B(q)\, q\, dq\, \{-(n_q + 1)\, f(\mathbf{p})[1 - f(\mathbf{p} - \mathbf{q})] + n_q f(\mathbf{p} - \mathbf{q})[1 - f(\mathbf{p})]\}.$$

(117)

The limits of integration over q are determined in this case from Eq. (115), assuming that $-1 \le \cos\theta \le 1$. The upper limit corresponds to $\cos\theta = 1$ and is equal to

$$q_2^{\text{ac}} = 2p - 2mu, \left.\begin{array}{c}\\ \\\end{array}\right\}$$
$$q_2^{\text{op}} = p + \sqrt{p^2 - 2m\hbar\omega_0.}$$

(118)

As before, for $q_2 > \hbar K$, the upper limit is equal to $\hbar K$, the lower limit q_1 also corresponds to $\cos\theta = 1$ (except for the case of acoustic phonons, when $\hbar\omega_q \to 0$ as $q \to 0$).

$$q_1^{\text{ac}} = 0, \left.\begin{array}{c}\\ \\\end{array}\right\}$$
$$q_1^{\text{op}} = p - \sqrt{p^2 - 2m\hbar\omega_0.}$$

(119)

Now the kinetic equation for the distribution function assumes the form

$$\frac{\partial f(\mathbf{p})}{\partial t} = \frac{Vm}{2\pi\hbar^4 p}\left\{\int^{+} B(q)\, q\, dq\, \{-n_q f(\mathbf{p})[1 - f(\mathbf{p} + \mathbf{q})] + \right.$$
$$+ (n_q + 1)\, f(\mathbf{p} + \mathbf{q})[1 - f(\mathbf{p})]\} + \int^{-} B(q)\, q\, dq\, \{-(n_q + 1)\, f(\mathbf{p})[1 - f(\mathbf{p} - \mathbf{q})] +$$
$$\left. + n_q f(\mathbf{p} - \mathbf{q})[1 - f(\mathbf{p})]\}\right\}.$$

(120)

For brevity, we have used the plus sign to indicate those integrals in which the upper limit is determined by Eq. (112), the minus sign for those in which the upper limit is determined by Eq. (118).

We investigate the retardation of electrons in the absence of an electric field. The momentum distribution of the carriers is isotropic, hence the distribution function depends only on the energy of the carriers, i.e., $f(\mathbf{p}) \equiv f(\varepsilon)$.

For further simplification of the kinetic equation, we make use of an expansion, first proposed in [76], of the integral in a series with respect to a small parameter. The small parameter in our case is the ratio of the energy of the emitted phonon to the electron energy. For scattering by the optical vibrations of the lattice, smallness of the parameter implies $\hbar\omega_0 < \varepsilon$, i.e., the energy of the electron must exceed the Debye temperature kT_D. The scattering of electrons by acoustic vibrations of the lattice, according to the conservation laws, is ascribed to phonons whose momentum is more than twice the momentum of the electron. Then for the ratio $\hbar\omega_q/\varepsilon$ we obtain $\hbar\omega_q/\varepsilon \sim 2u/v$, where v is the electron velocity. This ratio is less than unity to velocities of 10^6 cm $\cdot$ sec^{-1}, i.e., to energies of 10^{-3} eV (T < 10°K).

Now, expanding the distribution function in the integrand of the expression (120) and describing terms through $\hbar\omega_q$ squared, after a certain rearrangement of terms we obtain

$$
\begin{aligned}
\mathcal{I} = \int^{+} B\, qdq\, \{-n_q f(\mathbf{p})[1 - f(\mathbf{p}+\mathbf{q})] + (n_q + 1)f(\mathbf{p}+\mathbf{q})[1 - f(\mathbf{p})]\} + \\
+ \int^{-} Bqdq\, \{-(n_q + 1)f(\mathbf{p})[1 - f(\mathbf{p}-\mathbf{q})] + n_q f(\mathbf{p}-\mathbf{q})[1 - f(\mathbf{p})]\} = \\
= f(\varepsilon)[1 - f(\varepsilon)]\left(\int^{+} Bqdq - \int^{-} Bqdq\right) + f'(\varepsilon)[1 - f(\varepsilon)]\int^{+} B\hbar\omega_q q\, dq - f'(\varepsilon)f(\varepsilon)\int^{-} B\hbar\omega_q qdq + \\
+ f'(\varepsilon)\left(\int^{+} B\hbar\omega_q n_q qdq - \int^{-} B\hbar\omega_q n_q q\, dq\right) + \tfrac{1}{2}\left\{f''(\varepsilon)\left[\int^{+} Bq(\hbar\omega_q)^2 n_q dq + \int^{-} B(\hbar\omega_q)^2 n_q qdq\right] + \right. \\
\left. + f''(\varepsilon)[1 - f(\varepsilon)]\int^{+} B(\hbar\omega_q)^2\, qdq + f''(\varepsilon)f(\varepsilon)\int^{-} B(\hbar\omega_q)^2\, qdq\right\}.
\end{aligned} \tag{121}
$$

We now expand the integrals in powers of the energy of the emitted phonon. As already noted earlier, in the case of interaction with acoustic phonons, the lower limit of integration is always equal to zero. It is shown in detail in [55] why we neglect only third-order small terms with respect to the phonon energy for interaction with optical phonons when the lower limit is replaced by zero; a similar result is inferred by taking the upper limit correct to first-order smallness with respect to the phonon energy. Consequently, instead of Eqs. (112), (113), (118), and (119), we write

$$
\left.
\begin{aligned}
q_1^{\pm} &= 0, \\
q_2^{\pm} &= 2p \pm \frac{m\hbar\omega_{2p}}{p}.
\end{aligned}
\right\} \tag{122}
$$

After expansion of the integrals, the individual sums in (121) reduce to the following form:

$$
f(1 - f)\left(\int^{+} Bqdq - \int^{-} Bqdq\right) = f(1 - f)\frac{\partial}{\partial\varepsilon}\int_0^{2p} B(q)\hbar\omega_q qdq,
$$

$$
f'(1 - f)\int^{+} B\hbar\omega_q qdq - f'f\int^{-} B\hbar\omega_q qdq + f'\left(\int^{+} Bq\hbar\omega_q n_q d_q - \int^{-} Bq\hbar\omega_q n_q dq\right) =
$$

$$
= [f(1 - f)]' \int_0^{2p} Bq\hbar\omega_q dq + \tfrac{1}{2}f'\frac{\partial}{\partial\varepsilon}\int_0^{2p} Bq(\hbar\omega_q)^2(2n_q + 1)\, dq,
$$

$$
\tfrac{1}{2}\left\{f''\left[\int^{+} Bq(\hbar\omega_q)^2 n_q dq + \int^{-} Bq(\hbar\omega_q)^2 n_q dq\right] + f''(1 - f)\int^{+} Bq(\hbar\omega_q)^2\, dq + \right.
$$

$$
\left. + f''f\int^{-} Bq(\hbar\omega_q)^2\, dq\right\} = \tfrac{1}{2}f''\int_0^{2p} B(\hbar\omega_q)^2(2n_q + 1)\, qdq.
$$

Combining these expressions, we obtain for the quantity $\mathcal{J}$ determined in (121):

$$\mathcal{J} = \frac{\partial}{\partial \varepsilon} \left[f(1-f) \int_0^{2p} B(p)\hbar\omega_q q\,dq + \frac{1}{2} f' \int_0^{2p} B(q)\hbar\omega_q (2n_q + 1)\,q\,dq \right]. \tag{123}$$

Finally, the kinetic equation (102) may be written as follows:

$$\frac{\partial f(\varepsilon)}{\partial t} = \frac{1}{p}\frac{\partial}{\partial \varepsilon}\left\{ G(\varepsilon)\left[f(\varepsilon)(1-f(\varepsilon)) + \eta(\varepsilon)\frac{\partial f(\varepsilon)}{\partial \varepsilon}\right]\right\}, \tag{124}$$

where, for convenience, we have introduced certain quantities defined in [29]:

$$G(\varepsilon) = \frac{Vm}{2\pi\hbar^4}\int_0^{2p} B(q)\hbar\omega_q q\,dq, \tag{125}$$

$$G(\varepsilon)\,\eta(\varepsilon) = \frac{Vm}{4\pi\hbar^4}\int_0^{2p} B(q)(\hbar\omega_q)^2 (2n_q + 1)\,q\,dq. \tag{126}$$

In the absence of degeneracy, Eq. (124) goes over to equations derived earlier in [26, 28, 29, 76]. The collision time τ of a current carrier with a phonon is determined by the equation

$$\tau^{-1} = \frac{Vm}{4\pi\hbar^4 p^3}\int_0^{2p} B(q)(2n_q + 1)\,q^3\,dq. \tag{127}$$

The mean free path of the electron is equal to $l = v\tau$, where v is the electron velocity.

2. Determination of the Slowing-Down Time of Nonequilibrium Current Carriers

With the help of the kinetic equation (124), we readily obtain an equation for the variation of the mean energy of the electron gas, i.e., an equation describing the slowing-down process for electrons in a crystal. Recalling that the density of states of electrons with energy ε is equal to

$$\rho(\varepsilon) = 4\pi(2\pi\hbar)^{-3} V(2m)^{3/2}\varepsilon^{1/2}, \tag{128}$$

we multiply Eq. (124) by $\varepsilon\rho(\varepsilon)$. Integrating it over all values of the electron energy, we obtain

$$\frac{dE}{dt} = -\int_0^\infty \frac{\rho(\varepsilon)}{p} G(\varepsilon)\left[f(1-f) + \eta(\varepsilon)\frac{df}{d\varepsilon}\right]d\varepsilon, \tag{129}$$

where $E = \int \varepsilon\rho(\varepsilon)f(\varepsilon)d\varepsilon$ is the energy of the electron gas. The first term in the integral of Eq. (129) describes the loss of electron energy due to the spontaneous emission of phonons. This is the term that describes the slowing-down of the electrons, since an electron loses energy only as the result of spontaneous transitions, whereas the electron energy does not change on the average for induced transitions, which are proportional to the number of phonons n_q.

The spontaneous emission time t for one phonon is related to the relaxation time τ (collision time with a phonon) in order of magnitude by the relation

$$t \sim \tau \; (2n_q + 1). \tag{130}$$

The second term in Eq. (129) accounts for the process of phonon absorption by electrons and the process of induced phonon emission. It is apparent that if $f' > 0$, energy will be lost by the electrons; if $f' < 0$, the electrons will acquire energy from the lattice. For a slight departure from thermodynamic equilibrium, $f' < 0$.

The heating of electrons due to the lattice becomes substantial only when the temperature of the electron gas is near the lattice temperature. As will be apparent presently, inclusion of the second term in the integral of Eq. (129) for the scattering of electrons by acoustic vibrations of the lattice results in a log-infinite settling time for equilibrium between the electrons and the lattice.

At a lattice temperature equal to zero, inclusion of the second term in (129) means that further slowing down becomes impossible for certain electron energies due to the conservation of energy and momentum ("cold" electron [77]). In this case, however, the kinetic equation (129) is pushed to the limit of its interval of applicability, since the parameter of the expansion approaches unity.

For now, we will omit the second term in Eq. (129), confining our discussion to the retardation of electrons to energies exceeding the lattice temperature.

In the case of slowing down of an individual electron without regard for degeneracy,

$$f(\varepsilon)\,\rho(\varepsilon) = \delta(\varepsilon - E). \tag{131}$$

For interaction with acoustic phonons, the square of the matrix element, according to [78], has the form

$$B(q) = \frac{2}{9} \frac{C^2 q}{M n_0 u V}, \tag{132}$$

where C is the constant of interaction with acoustic phonons, M is the mass of the lattice atom, and n_0 is the density of lattice atoms.

Substituting (132) into (125), and integrating over q, we obtain

$$G(\varepsilon) = \frac{16 m^3 C^2 \varepsilon^2}{9 \pi \hbar^4 M n_0}. \tag{133}$$

The electron relaxation time in the case when $n_q \sim kT/\hbar\omega_q$ is equal to

$$\tau^{-1} = \frac{4 C^2 m^2 k T v}{9 \pi \hbar^4 M n_0 u^3}. \tag{134}$$

Substituting (131) and (133) into (129), we obtain an equation for the slowing down of an electron in interaction with acoustic vibrations of the lattice:

$$\frac{dE}{dt} = - a_0 E^{3/2}, \tag{135}$$

where

$$a_0 = \frac{2 \, (2m)^{1/2} \, u^2}{l_{ac}(T) \, kT}, \tag{136}$$

l_{ac} is the mean free path of the carrier when scattered by acoustic vibrations of the lattice. Inasmuch as $l \sim 1/T$, the value of the denominator in (136) does not depend on the temperature. The quantity a_0 is also conveniently expressed in terms of the electron mobility w(T) for scattering by acoustic vibrations of the lattice:

$$a_0 = \frac{8u^2 e}{3 \sqrt{\pi} w(T)(kT)^{3/2}}. \tag{137}$$

Integrating Eq. (135) from the initial electron energy E_0 to its final energy E, we obtain

$$t = \frac{2}{a_0}(E^{-1/2} - E_0^{-1/2}). \tag{138}$$

Thus, the slowing-down time t is governed chiefly by the final energy of the electrons E. This bears on the fact that the relaxation time (134) increases and the mean energy of the emitted phonons decreases with a reduction in the mean energy of the electron. Therefore, the region of small values of the energy leads to the longest slowing-down times.

We observe that the slowing-down time t depends on two constants of the semiconductor: the mobility and the velocity of sound. The ratio of the slowing-down times from the energy E_0 to E for $E_0 \gg E$ in two different semiconductors x and y is equal to

$$\frac{t_x}{t_y} = \frac{w_x u_y^2}{w_y u_x^2}. \tag{139}$$

Numerical estimates made for electrons in germanium (w $\sim$ 3600 cm$^2 \cdot$ W$^{-1} \cdot$ sec^{-1} at T = 300°K, u = 4.94 $\cdot$ 10^5 cm $\cdot$ sec^{-1}) for slowing down to an energy corresponding to 300°K yield t $\sim$ 4.5 $\cdot$ 10^{-10} sec. For electrons in silicon (w $\sim$ 1200 cm$^2 \cdot$ W$^{-1} \cdot$ sec^{-1}, u $\sim$ 8.5 $\cdot$ 10^5 cm $\cdot$ sec^{-1}), t $\sim$ 5.1 $\cdot$ 10^{-11} sec.

We now investigate the slowing down of a fast electron by optical vibrations of the lattice of an n-type semiconductor. In this case, Eqs. (124) and (129) are valid when the energy of the electron $E \gg \hbar\omega_0$, where $\hbar\omega_0$ is the energy of the optical phonon ($\hbar\omega_0 \approx kT_D$, where T_D is the Debye temperature).

The square of the interaction matrix element in this case has the form [78]

$$B(q) = \frac{K^2 D^2 \hbar^2}{2n_0 M V \hbar\omega_0}, \tag{140}$$

where K is the reciprocal lattice vector, D^2 is the constant of interaction with optical vibrations. Substituting (140) into (125), we obtain

$$G(\varepsilon) = \frac{D^2 K^2 m^2 \varepsilon}{\pi \hbar^2 M n_0}. \tag{141}$$

The relaxation time is equal to

$$\tau^{-1} = \frac{D^2 K^2 m^2 v}{2\pi \hbar^2 M n_0 \hbar\omega_0}. \tag{142}$$

We have regarded the lattice temperature as small in comparison with the Debye temperature.

Making use of the expressions obtained above, we arrive at an equation for determining the slowing-down time in interaction with optical phonons:

$$\frac{dE}{dt} = -\left(\frac{K^2 D^2 m^{3/2}}{\sqrt{2}\pi n_0 M \hbar^2}\right) E^{1/2}. \tag{143}$$

Integrating this equation, we find the slowing-down time from the initial energy E_0 to the final energy E:

$$t_{op} = \frac{2^{3/2}\pi M n_0 \hbar^2}{K^2 D^2 m^{3/2}} (E_0^{1/2} - E^{1/2}).$$ (144)

A comparison of the slowing-down times for optical and acoustic vibrations of the lattice yields

$$\frac{t_{op}}{t_{ac}} = \frac{16 C^2 m u^2}{9 D^2 \hbar^2 K^2 u^2}.$$ (145)

If we assume that slowing down occurs from an initial energy E_0 several times the Debye temperature to a final energy E near the latter, then, recognizing that $\hbar u K \approx k T_D$ in order of magnitude, we obtain

$$\frac{t_{op}}{t_{ac}} \sim \frac{C^2}{D^2} \frac{m u^2}{k T_D}.$$ (146)

For $C^2 \approx D^2$, the ratio of slowing-down times amounts to about 10^{-2}.

As indicated earlier, degeneracy of the nonequilibrium current carriers in the corresponding bands is essential to all the methods for obtaining negative temperature state utilizing direct transitions. It is instructive, therefore, to determine the slowing-down time under conditions of carrier degeneracy. Since degeneracy occurs for very large carrier densities in semiconductors, the smallest characteristic time in this case is the collision time of an electron with other electrons. As shown in [25], the time for the transfer of energy between electrons is equal in order of magnitude to

$$\tau_{ee} \sim \frac{\varepsilon^{3/2} m^{1/2} D^2}{\pi e^4 n_e},$$ (147)

where D is the dielectric constant, ε is the carrier energy, and n_e is the electron concentration.

Already at concentrations $n_e \sim 10^{14}$ cm^{-3}, the exchange of energy in electron collisions becomes more intense than in collision with acoustic phonons (for $T < T_D$ and $\bar{\varepsilon} < k T_D$, interaction with optical phonons is inconsequential). However, collisions between electrons does not lead directly to slowing down of the electron gas; rather it only results in the establishment of thermodynamic equilibrium inside the electron gas with a temperature Θ determining the total energy of this gas.

Consequently, at the electron densities required for the formation of negative temperature states in semiconductors, the form of the distribution function may be assumed known:

$$f(\varepsilon) = (1 + e^{(\varepsilon-\mu)/k\Theta})^{-1},$$ (148)

where μ is the Fermi quasi-level.

The slowing-down process amounts to a drop in the temperature Θ to the lattice temperature T. Earlier, the slowing-down time of electrons was determined without regard for degeneracy or the second term on the right side of Eq. (129). Now from Eq. (129) we obtain the time for the temperature of the electron gas to change from Θ_0 to Θ, considering only interaction with acoustic phonons, since allowance for degeneracy is important at low crystal temperatures and relatively small electron energies.

If the energy of the emitted phonons is less than kT, the parameter $\eta(\varepsilon)$ determined by Eq. (126) is equal to kT. Using the value of $G(\varepsilon)$ for the acoustic phonons (133) from Eq. (129), we obtain

$$\frac{dE}{dt} = -\frac{16m^3C^2k}{9\pi\hbar^4Mn_0}\int_0^\infty \frac{\varepsilon^2\rho(\varepsilon)}{p}(T-\Theta)\frac{\partial f}{\partial\varepsilon}\,d\varepsilon. \tag{149}$$

We made use of the relation

$$\frac{df}{d\varepsilon} = -\frac{1}{k\Theta}f(\varepsilon)[1-f(\varepsilon)]. \tag{150}$$

Integrating (149) by parts and recognizing that $f(\varepsilon)\to 0$ as $\varepsilon\to\infty$, we obtain

$$\frac{dE}{dt} = \frac{32m^{3/2}C^2k}{9\pi\sqrt{2}\hbar^4Mn_0}(T-\Theta)\int_0^\infty \varepsilon^{1/2}\rho(\varepsilon)f(\varepsilon)\,d\varepsilon. \tag{151}$$

Making use of the expression (136), the latter equation is rewritten

$$\left.\begin{array}{l} dE/dt = 2a_0k(T-\Theta)\overline{E^{1/2}}, \\[2mm] \overline{E^{1/2}} = \displaystyle\int_0^\infty \varepsilon^{1/2}\rho(\varepsilon)f(\varepsilon)\,d\varepsilon. \end{array}\right\} \tag{152}$$

Equation (152) is a differential equation for $\Theta(t)$, which can be reduced to an integral equation in elementary functions for sufficiently strong degeneracy, when it is permissible to assume $\mu/k\Theta \gg 1$. In this case, it is permissible to use an expansion of the integrals in (152) in powers of the temperature of the electron gas. Applying the customary expansion procedure [79] for the integrals E, μ, and $E^{1/2}$, we obtain

$$E = \frac{4\pi(2m)^{3/2}V}{(2\pi\hbar)^3}\frac{2}{5}\mu_0^{5/2}\left(1+\frac{5\pi^2}{12}\frac{k^2\Theta^2}{\mu_0^2}\right), \tag{153}$$

$$\overline{E^{1/2}} = \frac{4\pi(2m)^{3/2}V}{(2\pi\hbar)^3}\frac{1}{2}\mu_0^2\left(1+\frac{\pi^2}{6}\frac{k^2\Theta^2}{\mu_0^2}\right), \tag{154}$$

$$\mu = \mu_0\left(1-\frac{\pi^2}{12}\frac{k^2\Theta^2}{\mu_0^2}\right), \tag{155}$$

where $\mu_0 = \frac{(\pi\hbar)^2}{2m}\left(\frac{3n}{\pi}\right)^{2/3}$ is the chemical potential of the degenerate electron gas at the temperature $\Theta = 0$. If we substitute Eqs. (153) and (154) into (152), we find

$$\Theta\frac{\partial\Theta}{\partial t} = \frac{3a_0}{k\pi^2\mu_0^{1/2}}(T-\Theta)\left(\mu_0^2 + \frac{1}{6}\pi^2k^2\Theta^2\right), \tag{156}$$

whence the time for the temperature to vary from the initial Θ_0 to the final value Θ is equal to

$$t = \int_{\Theta_0}^{\Theta}\frac{k\pi^2}{3a_0\mu_0^{3/2}}\frac{\Theta\,d\Theta}{(T-\Theta)\left(1+\frac{\pi^2}{6}\frac{k^2\Theta^2}{\mu_0^2}\right)}. \tag{157}$$

This equation is integrated in elementary functions:

$$t = \frac{k\pi^2}{3a_0\mu_0^{3/2}}\left(1+\frac{\pi^2k^2T^2}{6\mu_0^2}\right)^{-1}\left\{T\ln\frac{(\Theta_0-T)\left(1+\frac{\pi^2k^2\Theta^2}{6\mu_0^2}\right)^{1/2}}{(\Theta-T)\left(1+\frac{\pi^2k^2\Theta_0^2}{6\mu_0^2}\right)^{1/2}} + \frac{\sqrt{6}\mu_0}{\pi k}\left[\mathrm{arctg}\left(\frac{\pi k\Theta_0}{\sqrt{6}\mu_0}\right) - \mathrm{arctg}\left(\frac{\pi k\Theta}{\sqrt{6}\mu_0}\right)\right]\right\}. \tag{158}$$

For $\Theta \to T$, the first term inside the braces in Eq. (158), which arises due to inclusion of the second term on the right side of (129), goes to infinity logarithmically, thus corresponding to an infinite stabilization time for equilibrium between the electrons and the lattice. In the case $\Theta > T$, the logarithm in (158) may be neglected, whereupon the slowing-down time has the following form upon expansion of the functions in powers of the temperature of the electron gas:

$$t \approx \frac{k\pi^2}{3a_0\mu_0^{3/2}} (\Theta_0 - \Theta) \tag{159}$$

(we are neglecting second-order terms).

The temperature of the electron gas involved in Eq. (159) may be expressed in terms of the average electron energy. We assume that the lattice temperature is zero. In this case, the final temperature of the electron gas may be assumed equal to zero and the electron energy equal to the average degeneracy energy. Consequently, we have

$$\left. \begin{aligned} \varepsilon &= {}^3/_5\mu_0, \\ \varepsilon_0 &= {}^3/_5\mu_0 \left(1 + \frac{5\pi^2}{12} \frac{k^2\Theta^2}{\mu_0^2} \right). \end{aligned} \right\} \tag{160}$$

Making use of (160), we obtain from (159)

$$t \approx \frac{2\pi}{5a_0\varepsilon^{1/2}} \left(\frac{\varepsilon_0}{\varepsilon} - 1 \right)^{1/2}. \tag{161}$$

Let us compare this equation with the equation for the slowing-down time for acoustic phonons (138), which was derived without regard for degeneracy of the electrons:

$$t \approx \frac{2}{a_0\varepsilon^{1/2}} \left(1 - \frac{\varepsilon^{1/2}}{\varepsilon_0^{1/2}} \right). \tag{162}$$

It is evident from the comparison that the values obtained for the slowing-down time with and without regard for the slowing time agree in order of magnitude, although the retardation law has a substantially different form in either case.

The fact that allowance for degeneracy does not lengthen the slowing-down time of the current carriers has a simple physical interpretation. The number of carriers ΔN for which the probability of phonon emission is not appreciably reduced by degeneracy is determined by the "flattening" of the Fermi distribution near the chemical potential. This "flattening" is of the order kT. Consequently, we can write $\Delta N \sim \Theta$. On the other hand, the excess ΔE of the kinetic energy over the degeneracy energy in connection with this "flattening" (only this part of the energy may be imparted to the lattice) is proportional to the number of particles ΔN and the magnitude of the "flattening" $k\Theta$. Therefore, $\Delta E \sim \Theta^2$. Hence it is apparent that as $\Theta \to 0$, the energy of the electron gas ΔE tends to zero more rapidly than the number of particles capable of effectively emitting phonons. This fact results in the slowing-down time remaining finite with allowance for degeneracy.

We point out that there is no need to consider the slowing down of the degenerate electron gas at optical phonons, because, in the cases of interest, the mean energy of the electrons for which degeneracy is significant is less than $\hbar\omega_0$.

Consequently, the slowing down of an electron gas in a semiconductor falls into two separate stages: slowing down at optical vibrations of the lattice to energies $\sim\hbar\omega_0$ with short characteristic times, according to Eq. (144), and further retardation at acoustic vibrations with much longer characteristic times according to (138) and (161). Consequently, the times for formation of negative temperatures are determined primarily by the slowing down of electrons from energies $\hbar\omega_0$ to energies corresponding to the necessary degree of degeneracy, taking into account interaction only with acoustic phonons.

CHAPTER VI

ENERGY DISTRIBUTION OF CURRENT CARRIERS IN SEMICONDUCTORS IN THE PRESENCE OF FAST ELECTRON SOURCES

At the present time, powerful light sources, including monochromatic laser radiation [62] and monoenergetic electron beams, are widely used for the excitation of electron−hole pairs in semiconductors. For a number of practical problems, especially for obtaining negative temperature states either between levels of one band or between levels of different bands, it is important to know the energy distribution of the nonequilibrium carriers formed by the creation in a semiconductor of electrons and holes whose energy at the instant of formation exceeds the lattice temperature of the crystal by a factor of tens or hundreds [8]. These nonequilibrium carriers† can be generated by an electron beam with energies of tens to hundreds of kiloelectronvolts or by light sources (including laser radiation), if the energy of the exciting quanta exceeds the width of the forbidden band. In our ensuing discussion, we will not be specific about the form of the fast carrier source; we will assume instead a prescribed initial energy distribution of the carriers created by the external source.

The kinetic equation for the nonequilibrium part‡ of the electron distribution function§ $f(\mathbf{p})$ in the conduction band, in the case when the carrier densities are still not so great that degeneracy or collisions between carriers is significant, has the form

$$(a - b) f(\mathbf{p}) + j(\mathbf{p}) - U(\mathbf{p}) = 0. \tag{163}$$

Here $(a - b) f(\mathbf{p})$ is the collision integral, including the interaction of electrons with the crystal lattice, $j(\mathbf{p})$ is the number of electrons per cm^3 with momentum $\mathbf{p}$ that are created in the conduction band by the external source per unit time, and $U(\mathbf{p})$ is the number of electrons per cm^3 with momentum $\mathbf{p}$ that leave the conduction band per unit time due to various recombination processes.

Clearly, $j(\mathbf{p})$ does not depend on the distribution function $f(\mathbf{p})$, whereas $U(\mathbf{p})$ does depend on $f(\mathbf{p})$. This dependence will be discussed a little later.

Let us assume that the term $U(\mathbf{p})$ is small in comparison with the first term in (163), when the energy $\varepsilon(\mathbf{p}) > kT$, where T is the lattice temperature. This means physically that the collision time of an electron with the lattice is less for energies $\varepsilon > kT$ than the lifetime with respect to recombination. This condition is always satisfied down to very low temperatures, which we will not consider. Then, in the region $\varepsilon > kT$, Eq. (163) has the following form for determining the form of the function $f(\mathbf{p})$ [80]:

$$(a - b) f(\mathbf{p}) + j(\mathbf{p}) = 0. \tag{164}$$

---

† We will refer to such carriers everywhere below as fast carriers.
‡ In the case when the nonequilibrium concentration is small in comparison with the thermodynamic-equilibrium concentration, clearly, the total function is the sum of the thermodynamic-equilibrium and the sought-after equilibrium functions.
§ All of the results, of course, are valid for holes of the valence band.

To write the collision integral in explicit form, we make use of the same expansion in the ratio of the energy of the interacting phonon to the electron energy $\hbar\omega_q / \varepsilon(\mathbf{p})$ as a small parameter, as we used in the derivation of the kinetic equation in §1 of Chapter V. Since there is no external electric field, and since $j(\mathbf{p})$ is an isotropic function of the momenta, i.e., $j(\mathbf{p}) \equiv j(\varepsilon)$, the collision integral is written only for the function $f(\varepsilon)$. If, as before, we let $\varepsilon(\mathbf{p}) = \mathbf{p}^2/2m$, where m is the effective mass of the carrier, then, according to the results of §1 of Chapter V, Eq. (164) may be written in the form

$$\frac{1}{p}\frac{\partial}{\partial\varepsilon}\left\{G(\varepsilon)\left[f(\varepsilon) + \eta(\varepsilon)\frac{\partial f(\varepsilon)}{\partial\varepsilon}\right]\right\} = j(\varepsilon). \tag{165}$$

All the notation in (165) is the same as in Eq. (124). The factor $[1 - f(\varepsilon)]$ dropped out, because degeneracy was disallowed. We note that our expansion in $\hbar\omega/\varepsilon$ in the case of optical phonons requires that $\bar{\varepsilon} \gg \hbar\omega_0$. If the distribution function differs little from a Boltzmann distribution with the lattice temperature, it is necessary that $kT \gg \hbar\omega_0$. Therefore, we will confine our discussion formally to the high-temperature case in the investigation of optical phonons.

Multiplying Eq. (165) by $\rho(\varepsilon) = 4\pi(2\pi\hbar)^{-3}V(2m)^{3/2}\varepsilon^{1/2}$, and integrating over ε from ∞ to ε, we obtain

$$f(\varepsilon) + \eta\frac{\partial f(\varepsilon)}{\partial\varepsilon} = \frac{V\overline{2m\varepsilon}}{\rho(\varepsilon)G(\varepsilon)}\int_{+\infty}^{\varepsilon} j(\varepsilon)\rho(\varepsilon)\,d\varepsilon, \tag{166}$$

where for acoustic phonons,

$$G(\varepsilon) = \frac{4mu^2\varepsilon^2}{l_{ac}(T)kT}, \quad \eta = kT, \tag{167}$$

and for optical phonons,

$$G(\varepsilon) = \frac{2\hbar\omega_0\varepsilon}{l_{op}(2n_0 + 1)}, \quad \eta = \frac{\hbar\omega_0(2n_0 + 1)}{2}, \quad n_0 = (e^{\frac{\hbar\omega_0}{kT}} - 1)^{-1}. \tag{168}$$

Here l is the mean free path, u is the velocity of sound, $\hbar\omega_0$ is the energy of the optical phonon.

The solution of Eq. (166) may be represented as the sum of the general solution of the homogeneous equation and one particular solution of the inhomogeneous equation:

$$f(\varepsilon) = Ce^{-\frac{\varepsilon}{kT}} + \frac{e^{-\frac{\varepsilon}{kT}}}{kT}\int_{\varepsilon_0}^{\varepsilon}\frac{e^{\frac{x}{kT}}V\overline{2mx}\,dx}{\rho(x)G(x)}\int_{+\infty}^{x} j(y)\rho(y)\,dy. \tag{169}$$

We assume that the source yields electrons with an energy no greater than ε_0, i.e.,

$$j(\varepsilon) \equiv 0 \quad \text{for} \quad \varepsilon \geqslant \varepsilon_0. \tag{170}$$

The constant C in the distribution function (169) is determined as follows. We multiply Eq. (163) by $\rho(\varepsilon)d\varepsilon$ and integrate over ε from zero to infinity. The integral of the first term is equal to zero, since the collision integral does not alter the total number of particles. We obtain

$$\int_0^\infty U(\varepsilon)\rho(\varepsilon)\,d\varepsilon = \int_0^\infty j(\varepsilon)\rho(\varepsilon)\,d\varepsilon = Q, \tag{171}$$

where Q is the total number of fast electrons per cm^3 created by the external source per unit time. The resulting equation expresses the condition of equality between the number of electrons created per $cm^3 \cdot sec$ in the conduction band and the number of electrons emerging from the band due to recombination processes.

As noted earlier, mainly electrons with energies $\varepsilon \leq kT$ recombine. Consequently, on the left side of Eq. (171) the region of integration actually goes from zero to kT. In this region, however, our assumptions anent the derivation of the distribution functions (169) are invalid. This difficulty may be lifted by assuming that there is a "sink" of recombining electrons for $\varepsilon = kT$ [81], so that

$$\int_0^\infty U(\varepsilon)\,\rho(\varepsilon)\,d\varepsilon \simeq f(kT)\,\frac{1}{\tau_c}\int_0^{kT}\rho(\varepsilon)\,d\varepsilon = Q, \tag{172}$$

where τ_c is the lifetime of the nonequilibrium electrons in the conduction band with respect to the recombination process.

Substituting (169) into (172), we find the constant C, whereupon the completely defined distribution function has the form

$$f(\varepsilon) = \frac{Q\tau_c e^{-\frac{\varepsilon}{kT}}}{kT\rho(kT)}\left\{1 + \frac{kT\sqrt{2mkT}}{G_n(kT)\,\tau_c}\int_1^{\frac{\varepsilon}{kT}}\left[\frac{e^x}{x^n}\,\frac{\int_\infty^{kTx} j(y)\,\rho(y)\,dy}{Q}\right]dx\right\}. \tag{173}$$

In Eq. (173), $n = 2$ in the case of acoustic phonons, $n = 1$ for optical phonons [of course, the form of $G_n(kT)$ is also different, a fact which we emphasize by the subscript n].

As implied by (173), the distribution function comprises a sum of two terms: the first, an ordinary Boltzmann distribution, the second reflecting the presence of fast electrons due to the constant influence of the source $j(\varepsilon)$. The coefficient in the braces in front of the second term is equal in order of magnitude to $\sim(kT/4mu^2)\tau_{ac}(T)/\tau_c$ in the case of acoustic phonons, to $\sim(T/T_D)^2\tau_{op}/\tau_c$ in the case of optical phonons, where τ_{ac} and τ_{op} are the collision times with acoustic or optical phonons, respectively; T_D is the Debye temperature.

Consequently, this coefficient, insofar as the derivation of $f(\varepsilon)$ is correct, is always much less than unity, and the distribution (175) is not significantly different from a Boltzmann distribution except for $\varepsilon \gg kT$. For $\varepsilon > \varepsilon_0$, the distribution again goes over to a Boltzmann distribution.

For a more detailed analysis of the expression (173), particularly in connection with the problem of realizing negative temperature state, we assume the external source of fast electrons to be monochromatic, i.e.,

$$j(\varepsilon) = A\delta(\varepsilon - \varepsilon_0), \tag{174}$$

where A is independent of the energy. It is apparent that such a source creates optical conditions for the realization of negative temperature states between levels within the corresponding band. Substituting (174) into (173), we obtain the following distribution function for $\varepsilon \leq \varepsilon_0$:

$$f(\varepsilon) = \frac{Q\tau_c e^{-\frac{\varepsilon}{kT}}}{kT\rho(kT)}\left[1 + \frac{kT\sqrt{2mkT}}{G_n(kT)\,\tau_c}\int_1^{\frac{\varepsilon}{kT}}\frac{e^x}{x^n}\,dx\right]. \tag{175}$$

In the case $\varepsilon \geq \varepsilon_0$, the upper limit in the integral is equal to ε_0/kT. The function (175) is monotonically decreasing with increasing ε, has everywhere a negative derivative with respect to ε, where $df/d\varepsilon$ tends asymptotically to zero with increasing ε. Hence follows the conclusion that it is impossible in the stationary regime, irrespective of the dependence on the power of the acting monochromatic source Q, to obtain the necessary inverse distribution between any two levels of the conduction band (negative temperature state) required in order to amplify the monochromatic radiation. This result is evident from the following elementary considerations. In the stationary regime, the electron flux in energy space $S(\varepsilon)$ = const. On the other hand, $S(\varepsilon) \sim f(\varepsilon)\rho(\varepsilon)d\varepsilon/dt$, where $d\varepsilon/dt$ is the energy lost by an electron with energy ε per unit time (determined in §2 of Chapter V). From the condition $S(\varepsilon)$ = const we have $f(\varepsilon) \sim [\rho(\varepsilon)d\varepsilon/dt]^{-1}$. But $[(d\varepsilon/dt)\rho(\varepsilon)] \sim \varepsilon^n$, $n > 0$ (n = 2 for acoustic phonons, n = 1 for optical phonons), i.e., $f(\varepsilon)$ is a monotonically decreasing function.

We have not investigated the case when it is possible in the conduction band in a strong magnetic field for the Landau levels to occur for the mobile carriers between which it was proposed to create negative temperature states by the method of monochromatic pumping [6, 7]. However, there are no theoretical arguments that would alter the above result of the impossibility of an inverse distribution between these levels, especially in the case when the separation between neighboring Landau levels with a different quantum number does not exceed the optical phonon energy.

It is difficult to obtain an inverted population between levels belonging to the same band in the pulsed regime, since even after a time of 10^{-11} to 10^{-12} sec equilibrium is established between the electrons and phonons.

The above considerations indicate that it is possible with monochromatic pumping to obtain an inverse distribution in the stationary regime for levels in one band (ε_2 and ε_1, where $\varepsilon_2 > \varepsilon_1$), provided the following inequality is fulfilled:

$$\frac{d\varepsilon}{dt}\bigg|_{\varepsilon=\varepsilon_2} \rho(\varepsilon_2) < \frac{d\varepsilon}{dt}\bigg|_{\varepsilon=\varepsilon_1} \rho(\varepsilon_1),$$

i.e., the energy loss rate must increase with diminishing energy more rapidly than the density of levels decreases.

This case could be realized in principle under the following circumstances. The electron (hole) density is so large that the energy lost by a fast electron per unit time in collision with other electrons practically at the lattice temperature is greater than the power lost in the emission of phonons (mainly optical phonons), i.e., $(d\varepsilon/dt)_{ee} \gg (d\varepsilon/dt)_{ph}$. In this case, $f(\varepsilon) \sim \rho^{-1/2}(\varepsilon)\varepsilon^{1/2}$, and if $\rho(\varepsilon)$ diminishes with energy more slowly than $\varepsilon^{1/2}$, it is to be expected that for levels with energy $\varepsilon \gg kT$ the inverted population required for the amplification or generation of radiation will occur. As we know [25], satisfaction of the condition $(d\varepsilon/dt)_{ee} > (d\varepsilon/dt)_{ph}$ requires large electron (hole) densities in the conduction (valence) band (n > 10^{15} cm^{-3}), thus increasing considerably the absorption of the radiation for which an inverted population is obtained, and making it difficult to obtain a negative absorption coefficient.

CONCLUSION

Following our detailed investigation of the various methods for stimulating semiconductors as lasers, we are in a position to compare certain ones with others.

The method of nonequilibrium carrier injection through a p–n junction of degenerate semiconductors can, under certain conditions, yield an efficiency near unity in the transformation of electrical energy into coherent radiation. There are brief reports that already, today, this method of excitation has been used to attain efficiencies of 70 to 80%.

From the point of view of the most favorable and convenient transformation of input energy into the energy of radiated coherent light, therefore, this method is far more preferable than all the others. Moreover, a major advantage of the injection method is the simplicity of construction of such an oscillator, to the extent that devices may be fabricated with relatively small over-all dimensions. The use of degenerate semiconductors with forbidden bands of various widths makes it possible to achieve the generation of radiation by this method over a wide range of wavelengths, although, to be sure, not from all semiconductors is it possible to obtain degenerate p–n junctions with a low resistivity, permitting the elimination of overheating of the crystal at the required current densities. Right now, with the injection method, it has been possible to realize generation in the visible range. The rather low emitted power (tens of watts) attainable at the present time using this method of excitation is clearly attributable to the problem of cooling the p–n junction and the contiguous layers. It is anticipated that an increase in the efficiency and area of the p–n junction (which currently comprises a fraction of a square millimeter) will raise the emitted power.

The method of excitation by an electron beam, although its efficiency cannot be made higher than 40%, is capable of exciting large volumes of matter. It is hoped, therefore, to obtain high coherent radiation power outputs by means of an electron beam (for example, with 100-A current pulses and electron energies of 500 keV, a coherent radiation power of 10^7 W is expected). Excitation by an electron beam enables one to obtain negative temperature states for semiconductors with an energy width of the forbidden band, provided only that the slowing-down time of the current carriers is less than the lifetime. Excitation by primary electrons, therefore, will probably make it possible to generate in the shortest-wave region of the visible spectrum.

As noted earlier, the method of exciting semiconductors by means of laser beams gives the longest-wave radiation with inevitable energy losses. However, the use of laser radiation, especially from oscillators with a modulated quality factor, should prove useful in selecting suitable semiconductor materials for other modes of excitation. As implied by §1 of Chapter IV, the radiation from sufficiently powerful lasers, focused on a small surface of a semiconductor, can give such a high excitation density that the energy separation of the Fermi quasi-levels for electrons and holes can be made to attain the magnitude of the exciting quantum energy, which, given a forbidden band whose width is considerably less than the energy of the incident phonons, will result in a high degree of degeneracy of the nonequilibrium carriers (electrons and holes). It would be interesting to study the properties of such strongly nonequilibrium states.

The method of exciting homogeneous semiconductors with a strong electric field yields a low efficiency, since the available energy is expended in the time of one pulse as Joule heat and spontaneous recombination, a large current density with high voltages is required, and operation can only be realized in the pulsed mode. Clearly, with this method of excitation, it is difficult to build a laser.

An interesting approach to the design of an oscillator in semiconductors is the use of indirect transitions (from electrons and holes in the free or exciton states). In this case, it is required for the realization of negative temperature states to have sufficiently small nonequilibrium concentrations that operation in the continuous mode will be realized, while excitation is possible even using fairly powerful conventional light sources. If the energy of the emitted phonon is high, the oscillator can function also at not too low temperatures.

The present article has been devoted mainly to laying the groundwork for the potential creation of semiconductor quantum oscillators. The majority of the calculations were carried out prior to the time that coherent radiation was actually produced using semiconductors, at which time there was doubt as to the realizability of such a device. The creation of semiconductor quantum oscillators has given birth to a number of new theoretical and experimental problem areas associated with their operation. Among such problems is that of determining the mechanism of radiative transfer yielding coherent radiation (for example, the transfer producing a photon with an energy of 1.47 eV in GaAs). The solution of this problem is complicated at present by the fact that induced radiation is observed only with injection through a p−n junction of degenerate semiconductors, when the structure of the energy bands must be considered at large concentrations of impurities, interacting with each other and with the band.

It will be necessary for the further improvement of oscillators to ascertain the influence of various factors on the radiation degree of monochromaticity, which is rather low for existing semiconductor oscillators (the quality of the emission line is approximately 10^4).

It is very important to determine the dependence of the absorption coefficient (negative and positive) of the active medium in injection through a p−n junction on the current density. The solution of this problem requires knowledge of all the characteristic parameters of the p−n junction, nonradiative recombination processes, and processes involving the interaction of radiation with the current carriers. The solution of the problem of the spatial distribution of the injected carriers will be instrumental in more precisely calculating the efficiency and determining the types of electromagnetic modes that occur in the oscillator.

As inferred from the fundamental postulates of the present article, the realization of negative temperature states is only possible in the event of small carrier slowing-down times by comparison with their lifetime. Some theoretical calculations of the slowing-down times were carried out in the fifth chapter. However, so far it is impossible to make such calculations for the lifetimes at high nonequilibrium carrier concentrations, because we are still lacking any theory relating the lifetime in a semiconductor to other measurable constants. The solution of this problem will help in narrowing down the class of semiconductors suitable for the design of quantum oscillators. For certain methods of excitation (for example, the optical method), the active medium can only be formed near the surface of the semiconductor, which means that allowance for the influence of the surface on the lifetime acquires particular importance.

It is well known that the optical homogeneity of the active medium plays a major role in the generation of coherent radiation. The experimental and theoretical investigation of the various factors tending to deteriorate the optical homogeneity (role of impurities, defects, dislocations) is a new problem area in the physics of semiconductors.

In analyzing the various methods of excitation of semiconductors, we have postulated that the temperature of the crystal lattice remained constant. Of course, using very short pulses for excitation with adequate separation of the pulses, it is possible to avoid unwanted heating of the lattice. However, in the experiments that have been performed to date, the duration and power of the exciting pulses have been such that heating of the crystal lattice has proved the major detriment to the realization of negative temperature states. Only a detailed investigation of the process of cooling the excited portion of the semiconductor will permit the optimum constructions and operating conditions to be decided for the quantum oscillator.

Mention was also made of the potential value of a magnetic field in retuning the oscillator frequency and lowering the absorption by free carriers. It would be most interesting to carry out theoretical calculations by which the quantitative characteristics of these effects could be determined.

In the article, we indicated a number of other, as yet unsolved problems bearing on the application of semiconductors for the generation of coherent light (for example, the role of the "pinch effect" in the transmission of large currents through a homogeneous semiconductor, distribution of the current carriers formed as a result of the process of multistage ionization, etc.).

So far, it has only been possible to realize the generation of coherent light with semiconductors by injection of current carriers through a p−n junction. However, obviously, this stage was the most difficult in the problem of using semiconductors for quantum oscillators. It suffices to say that only a year has passed since the first report of an active semiconductor oscillator, and already generation has been realized by means of the semiconductor compounds GaAs, InP, InAs, and solid solutions of GaAs and GaP, InAs and GaAs. The number of theoretical investigations in this area, of diverse new hypotheses for specific models of quantum oscillators utilizing semiconductor materials, has burgeoned rapidly. Even though semiconductor quantum oscillators are still in their infant stage of development, the new possibilities that they afford in the generation of coherent light will undoubtedly assure them a solid future in science and technology.

The author would like to conclude with heartfelt appreciation to Corresponding Member of the Academy of Sciences of the USSR, N. G. Basov and Master of Physicomathematical Sciences, O. N. Krokhin, for extremely useful discussions regarding the problems covered by the present article.

APPENDIX I

CALCULATION OF THE TRIPLE RECOMBINATION CROSS SECTION IN SEMICONDUCTORS

It was indicated in §3 of Chapter I that the principal factor for large concentrations of current carriers in semiconductors could be the process of triple recombination, which is the converse of the process of impact ionization. Since the conservation of momentum prohibits the recombination of a slow electron and hole with the transfer of energy to a third slow carrier (see §2 of Chapter IV, where proof is given of the impossibility of a pair being created by an electron whose kinetic energy is less than the width of the forbidden band), impurities must contribute largely to the processes involved in their recombination.

The recombination cross sections of such processes were first calculated in [83]. In that paper, however, the orthogonality of the wave functions describing the motion of electrons with different quasi-momenta and energies was not systematically accounted for, as a result of which far too excessive values were obtained in some cases for the cross sections. These cross sections were calculated somewhat differently in [34], where it was also indicated that the proper calculation does not give anomalously large cross sections.

Using the first approximation of perturbation theory, we wish to calculate the cross section for capture of a free carrier by a vacant (for that carrier) impurity level when the carrier in question encounters all other slow carriers. In this case, the impurity can acquire any momentum, and in the capture of the carrier by an impurity level, energy will be imparted to the other carrier. We will carry out the calculation in the Born approximation. This means that the wave functions for free carriers in the crystal are assumed to be in the form of plane waves with a Bloch factor, so that the normalized wave functions we use will have the following form:†

---

† We use energy units for the momentum, i.e., the usual momentum is our momentum divided by the velocity of light.

$$\Psi_e(\mathbf{p}_e, \mathbf{r}) = \frac{u_e(\mathbf{r})}{\sqrt{\bar{V}}} e^{\frac{i\mathbf{p}_e\mathbf{r}}{\hbar c}},$$

$$\Psi_b(\mathbf{r}) = \frac{u_b(\mathbf{r})}{(\pi a_l^3)^{1/2}} e^{-\frac{r}{a_l}}, \qquad\qquad\qquad (A.1)$$

$$\Psi_h(\mathbf{p}, \mathbf{r}) = \frac{u_h(\mathbf{pr})}{\sqrt{\bar{V}}} e^{\frac{i\mathbf{pr}}{\hbar c}},$$

where the functions $\Psi_e(\mathbf{p}_e, \mathbf{r})$ and $\Psi_h(\mathbf{p}, \mathbf{r})$ describe an electron in the conduction band or a hole in the valence band, respectively, with momentum $\mathbf{p}$, and $\Psi_b(\mathbf{r})$ describes an electron or hole captured by an impurity atom. We postulate a hydrogen-like model for the impurity atom, with an effective radius

$$a_l = \left(\frac{E_R}{E_c - E_T}\right)^{1/2} a_H, \qquad\qquad\qquad (A.2)$$

where a_H is the radius of the Bohr orbit, E_R is the Rydberg energy, and $E_c - E_T$ is the depth of the energy level of the impurity, measured from the corresponding band. For the energy of interaction we use the Coulomb energy of interaction, taking into account Debye screening at a distance

$$L_D = \frac{1}{\sqrt{2}}\left(\frac{D\hbar^2}{\pi N m^* e^2}\right)^{1/4}, \qquad\qquad\qquad (A.3)$$

where D is the dielectric constant of the semiconductor, N is the free carrier concentration, and m* is the effective mass of the carrier.

Consequently, the energy of interaction between two particles may be written in the form

$$H_i = \frac{e^2}{D|\mathbf{r}_1 - \mathbf{r}_2|} \exp\left(-\frac{\mathbf{r}_1 - \mathbf{r}_2}{L_D}\right). \qquad\qquad\qquad (A.4)$$

In the first approximation of perturbation theory, we calculate the matrix element for the process of capture of an electron with momentum $\mathbf{p}_{e1}$ by an unfilled impurity level on encountering an electron with momentum $\mathbf{p}_{e2}$. The wave function of the initial state has the form

$$\Psi_I(\mathbf{r}_1\mathbf{r}_2) = \frac{1}{\sqrt{2}}[\Psi_e(\mathbf{p}_{e1}\mathbf{r}_1)\Psi_e(\mathbf{p}_{e2}\mathbf{r}_2) \pm \Psi_e(\mathbf{p}_{e1}\mathbf{r}_2)\Psi_e(\mathbf{p}_{e2}\mathbf{r}_1)], \qquad\qquad (A.5)$$

where the plus or minus sign depends on whether the electron has antiparallel spins (singlet state) or parallel spins (triplet state).

The wave function of the final state has the form

$$\Psi_{II}(\mathbf{r}_1\mathbf{r}_2) = \frac{1}{\sqrt{2}}[\Psi_e(\mathbf{p}\mathbf{r}_1)\Psi_b(\mathbf{r}_2) \pm \Psi_e(\mathbf{p}\mathbf{r}_2)\Psi_b(\mathbf{r}_1)], \qquad\qquad (A.6)$$

where $\mathbf{p}$ is the momentum of an electron acquiring the energy transmitted by another electron when it is implanted in an impurity atom, such that

$$\frac{\mathbf{p}^2}{2m^*c^2} = \frac{\mathbf{p}_{e1}^2}{2m^*c^2} + \frac{\mathbf{p}_{e2}^2}{2m^*c^2} + (E_c - E_T). \qquad\qquad\qquad (A.7)$$

The matrix element corresponding to transition from state I to state II has the form

$$(\Psi_{II}|H_i|\Psi_I) = \iint \frac{e^2 \exp\left(-\frac{|\mathbf{r}_1 - \mathbf{r}_2|}{L_D}\right)}{D(\mathbf{r}_1 - \mathbf{r}_2)} \Psi_{II}(\mathbf{r}_1\mathbf{r}_2)\Psi_I(\mathbf{r}_1\mathbf{r}_2)d_3r_1 d_3r_2. \qquad (A.8)$$

In the case of a singlet state, the matrix element (A.8) reduces to the following form:

$$(\Psi_{II} | H_i | \Psi_I) = \frac{e^2}{V^{3/2} (\pi a_l^3)^{1/2} D} (I_1 + I_2),$$
(A.9)

where

$$I_1 = \iint \frac{d_3 r_1 d_3 r_2}{|\mathbf{r}_1 - \mathbf{r}_2|} \exp \left(- \frac{|\mathbf{r}_1 - \mathbf{r}_2|}{L_D} - \frac{r_1}{a_l} - \frac{i}{\hbar c} \mathbf{p} \mathbf{r}_2 + \frac{i}{\hbar c} \mathbf{p}_0 \mathbf{r}_1 \right),$$
(A.10)

$$I_2 = \iint \frac{d_3 r_1 d_3 r_2}{|\mathbf{r}_1 - \mathbf{r}_2|} \exp \left(- \frac{|\mathbf{r}_1 - \mathbf{r}_2|}{L_D} - \frac{r_2}{a_l} - \frac{i}{\hbar c} \mathbf{p} \mathbf{r}_1 + \frac{i}{\hbar c} \mathbf{p}_0 \mathbf{r}_1 \right).$$
(A.11)

In the derivation of (A.9), we assumed $|\mathbf{p}_1, \mathbf{p}_2| \ll |\mathbf{p}|$, denoting by $\mathbf{p}_0$ either of the initial momenta $\mathbf{p}_{e1}$ or $\mathbf{p}_{e2}$. We also dropped the Bloch factors, which only give a correction on the order of unity. The change of variables $\mathbf{u} = (\mathbf{r}_2 + \mathbf{r}_1)/\sqrt{2}$ and $\mathbf{v} = (\mathbf{r}_2 - \mathbf{r}_1)/\sqrt{2}$ reduces the integral I_1 to the form

$$I_1 = \iint \frac{d_3 v d_3 u}{\sqrt{2} |\mathbf{v}|} \exp \frac{1}{\hbar c} \left(- \frac{\sqrt{2} |\mathbf{v}|}{L_D'} - i \sqrt{2} \mathbf{p} \mathbf{v} \right) \exp \frac{i}{\hbar c} \left[- \frac{|\mathbf{u} - \mathbf{v}|}{\sqrt{2} a_l'} - \frac{i}{\sqrt{2}} \mathbf{p} (\mathbf{u} - \mathbf{v}) \right],$$
(A.12)

where

$$L_D' = \frac{L_D}{\hbar c}, \qquad a_l' = \frac{a_l}{\hbar c}.$$

We again make the change of variables $\mathbf{v} = \mathbf{v}$, $\mathbf{z} = \mathbf{u} - \mathbf{v}$, whereupon

$$I_1 = \iint \frac{d_3 v d_3 z}{\sqrt{2} |\mathbf{v}|} \exp \frac{1}{\hbar c} \left(- \frac{\sqrt{2} |\mathbf{v}|}{L_D'} - i \sqrt{2} \mathbf{p} \mathbf{v} \right) \exp \frac{1}{\hbar c} \left(\frac{|\mathbf{z}|}{\sqrt{2} a_l'} - \frac{i}{\sqrt{2}} \mathbf{p} \mathbf{z} \right).$$
(A.13)

In a spherical coordinate system, I_1 has the form

$$I_1 = \frac{(2\pi)^2}{\sqrt{2}} \int_0^\infty v \, dv \int_{-1}^1 dx_1 \exp \left[\frac{v}{\hbar c} \left(- i \sqrt{2} p x_1 - \frac{\sqrt{2}}{L_D'} \right) \right] \times$$
$$\times \int_1^\infty z^2 dz \int_{-1}^1 \exp \left[\frac{z}{\sqrt{2} \hbar c} \left(- \frac{1}{a_l} + i |\mathbf{p}| x_2 \right) \right] dx_2.$$
(A.14)

Integrating, we obtain

$$I = \frac{2 (4\pi)^2 (\hbar c)^5 (\hbar c / a_l)}{\left[\left(\frac{\hbar c}{L_D} \right)^2 + \mathbf{p}^2 \right] \left[\left(\frac{\hbar c}{a_T} \right)^2 + \mathbf{p}^2 \right]^2}.$$
(A.15)

By suitable change of variables, the integral I_2 is brought to the form

$$I_2 = \frac{1}{\sqrt{2}} \iint \frac{d_3 v \, d_3 u}{|\mathbf{v}|} \exp \left[\frac{1}{\hbar c} \left(- \frac{\sqrt{2} |\mathbf{v}|}{L_D'} - i \sqrt{2} \mathbf{p} \mathbf{v} \right) \right] \exp \frac{1}{\hbar c} \left[\frac{(\mathbf{u} - \mathbf{v})}{\sqrt{2} a_l'} - \frac{i}{\sqrt{2}} \mathbf{p} (\mathbf{u} - \mathbf{v}) \right],$$
(A.16)

in other words, $I_2 \equiv I_1$.

Consequently, for the singlet state, we have

$$(\Psi_{II}^\bullet | H_i | \Psi_I) = \frac{2e^2}{V^{3/2} (\pi a_l^3)^{1/2}} 2 I_1,$$
(A.17)

where

$$I_1 = \frac{2\,(4\pi)^2\,L_D^2 a_t^3}{F\left(\dfrac{pL_D}{\hbar c}\right) F^2\left(\dfrac{pa_t}{\hbar c}\right)}, \qquad F\,(x) = \frac{1}{1+x^2}. \qquad (A.18)$$

The analogous calculation for the triplet state (antisymmetrical wave function) in the approximation $|\mathbf{p}_0| \ll |\mathbf{p}|$ yields zero. The cross section of the process is determined according to the usual equation [57]:

$$d\sigma = \frac{2\pi V}{\hbar v_0}\left(\frac{3}{4}\,|\,H_{tr}\,|^2 + \frac{1}{4}\,|\,H_{sin}\,|^2\right)\rho_f, \qquad (A.19)$$

where v_0 is the relative velocity of the colliding electrons, $|H_{tr}|^2$ and $|H_{sin}|^2$ are the squares of the matrix elements for transition in the triplet and singlet states. The density of the levels in the final state is

$$\rho_f = \frac{4\pi V p m c^2}{(2\pi \hbar c)^3}. \qquad (A.20)$$

Inasmuch as we are concerned with the capture of an electron by an impurity level on collision of this electron with any thermal electron, the total capture cross section is obtained by multiplication of the cross section defined by Eq. (A.19) by the number of all thermal electrons (where, of course, it is necessary to take the average thermal velocity $v_0 \cong \sqrt{kT/m^*}$). The cross section for capture of an electron by an impurity level on encountering all other thermal electrons is equal to

$$\sigma_{ee} = 4\pi^2\left(\frac{16\hbar^2 c^2 m^*}{p^2 a_H m D}\right)^2\left(\frac{p}{p_0}\right)(a_t^3 N_e)F^4\left(\frac{pa_t}{\hbar c}\right), \qquad (A.21)$$

where N_e is the density of electrons in the conduction band, m^* is the effective mass of the electron in the conduction band, and m is the ordinary mass of the electron. The inequality $pL_D \gg 1$ was used in the derivation of (A.21).

The cross section for capture of a hole by an electron-occupied impurity level on encountering other holes, clearly, is written in exactly the same form as σ_{ee}, except that the effective mass of the hole must be used and N_e must be replaced by N_h. Analogous cross sections are also obtained in the collision of an electron with holes, when capture of an electron by an impurity level occurs, while energy is transmitted to the hole, or when a hole is captured by an impurity level in collision with electrons. The anomalously large cross section was obtained in [83] for the process when a slow electron enters the valence band (recombining with a slow hole), while energy is transmitted to an impurity electron ejected into the conduction band. The essential fact is that in this case allowance was not made in [83] for the Bloch functions in Eq. (A.1), which ensure the orthogonality of the wave functions of the electron in different bands with the same quasi-momentum. Entirely analogous calculations made in the derivation of the expression for the matrix element (A.8) result in an integral $\overline{I}_1$ analogous to I_1:

$$\overline{I}_1 = \iint \frac{d_3 r_1 d_3 r_2}{|\mathbf{r}_1 - \mathbf{r}_2|}\,u_h^*\,(\mathbf{p}_0,\,\mathbf{r}_1)\,u_e\,(\mathbf{p}_0,\,\mathbf{r}_1)\,\exp\frac{1}{\hbar c}\left(-\,i\mathbf{p}\mathbf{r}_2 - \frac{|\mathbf{r}_1 - \mathbf{r}_2|}{L_D} - \frac{r_2}{a_t}\right). \qquad (A.22)$$

For simplicity, we assumed $\mathbf{p}_h = \mathbf{p}_e = \mathbf{p}_0$ and, for now, left out the Bloch functions.

If the product of the Bloch functions in the integral I_1 is assumed equal to unity, as was done in [83], then, clearly, $r_2 \sim a_t$ in the integral $\overline{I}_1$, whereas $|\mathbf{r}_1 - \mathbf{r}_2| \sim L_D$. In the integral I_1, on the other hand, $r_1 \sim a_t$, but $|\mathbf{r}_2 - \mathbf{r}_1| \sim \hbar c/p$ and, due to the fact that $L_D \gg \hbar c/p$, we also obtain for $\overline{I}_1$ a value much larger than for I_1. However, if we account systematically for the orthogonality of the Bloch functions, representing the integral as the sum of the integrals taken over all unit cells of the crystal and include the first and second terms of the expansion of the slowly varying function inside one cell, i.e., if

$$\Phi (R_i + r) \, u_h^* (p_0, R_i + r) \, u_e (p_0, R_i + r) = \left[\Phi (R_i) + \frac{\partial \Phi}{\partial (R_i + r)} \, r \right] u_h^* u_e, \qquad (A.23)$$

then the utilization of the orthogonality properties of the functions results in an additional small factor $\sim a / L_D$, where a is the dimension of the unit cell. Consequently, this factor squared (the cross section is proportional to the square of the matrix element) gives in order of magnitude the same values of the cross sections that we obtained previously.

In [84], an estimate was made of the same integral, and an analogous result was obtained. An order-of-magnitude estimate for $E_c - E_T = 0.3$ eV, $m^*/m = 0.3$, $\hbar c / p_0 = 1.7 \cdot 10^{-7}$ cm, $N_e \sim 3 \cdot 10^{13}$ cm^{-3} gives $\sigma_{ee} \sim 10^{-20}$ cm^2. However, as implied by the expression for σ_{ee} (A.21), with diminishing depth of the level $E_c - E_T$, the cross section increases in proportion to the third power of the diminution of this depth. Hence, if capture occurs at very shallow hydrogen-like levels of the impurities, as might be observed at low lattice temperatures, the cross section can become so large that such a process will play the most important role in the capture of carriers by impurities. Moreover, with a reduction in temperature, p_0 also decreases, thus increasing the cross section. At low temperatures, however, it is essential to refine the Born approximation we used in our calculations.

So far we have adopted the wave functions of the free carriers in the form of plane waves. Strictly speaking, this would be valid if we had neutral capture centers. However, in the case of charged centers, the Born approximation, as shown in [26], is valid in semiconductors if the initial energy ε_0 of the free carriers satisfies the inequality $\varepsilon_0 \gtrsim e^4 m^* / D^2 \hbar^2$. If we suppose that $\varepsilon_0 \sim kT$, we find that the Born approximation can be used to temperatures $T \gtrsim 100°K$ ($m^* \sim 0.1 \, m_e$, $D \sim 16$).

It is remarked in [84] that an estimate of the action of the Coulomb field in capture by a charged center in the case of triple recombination reduces the magnitude of the recombination coefficient only by a factor of $\frac{1}{1.5}$ to $\frac{1}{2}$ (by comparison with a neutral center with the same parameters), which generally falls within the limits of possible error in the framework of the given assumptions.

APPENDIX II

THE GAIN IN THE CASE OF DIRECT TRANSITIONS
IN SEMICONDUCTORS

Let us now calculate the gain in semiconductors, assuming that there are direct interband transitions and that the extrema of the conduction and valence bands occur at the electron quasi-momentum (wave vector) $k = 0$. The variation of the number N_q of quanta per unit volume with energy $\hbar\omega$ and wave vector q and polarization vector e_q due to transitions from the level i to the level f is determined by the following expression:

$$\frac{\partial N_q}{\partial t} = \int \frac{2\pi}{\hbar} | H_{if} |^2 [f_e (\varepsilon_i) + f_h (\varepsilon_f) - 1] \rho_f \delta (\varepsilon_i + \varepsilon_f + \Delta - \hbar\omega) \, d\varepsilon, \qquad (A.24)$$

where $|H_{if}|^2$ is the square of the matrix element of the interaction operator

$$H = - \frac{ie\hbar}{\sqrt{D} \, mc} A\nabla; \qquad (A.25)$$

A is the usual vector potential of the electromagnetic field (without regard for the propagation of light in the medium); D is the dielectric constant of the medium; $f_e(\varepsilon_i)$ is the distribution function of electrons, where ε_i is read upwards from the bottom of the conduction band; $f_h(\varepsilon_f)$ is the distribution function of holes, where ε_f is read downwards from the top of the valence band; ρ_f is the density of energy levels in the final state; and Δ is the width of the forbidden band.

The conservation of momentum is taken into account in Eq. (A.24), so that

$$\mathbf{k}_i = \mathbf{k}_f + \mathbf{q}. \tag{A.26}$$

Since $|\mathbf{k}_{i,f}| \gg |\mathbf{q}|$, the conservation of momentum yields $\mathbf{k}_i \approx \mathbf{k}_f$.

Writing the expression for the absorption coefficient α, we include in Eq. (A.24) only induced transitions, which are proportional to $N_\mathbf{q}$, so that

$$\alpha = \frac{V\sqrt{D}}{cN_\mathbf{q}} \frac{\partial N_\mathbf{q}}{\partial t} = \frac{4\,V\sqrt{2}e^2}{V\sqrt{D}\,m^2c\omega\hbar^3}|P_{if}|^2 \left(\frac{m_e m_h}{m_e + m_h}\right)^{3/2}(\hbar\omega - \Delta)^{1/2} \times$$
$$\times \left\{ f_e\left[\frac{m_h}{m_h + m_e}(\hbar\omega - \Delta)\right] + f_h\left[\frac{m_e}{m_e + m_h}(\hbar\omega - \Delta)\right] - 1\right\}, \tag{A.27}$$

where m_e and m_h are the effective masses of the electron and hole, respectively, and P_{if} is the matrix element of the momentum operator, taken over the unit cell of the crystal:

$$P_{if} = -i\hbar \int \psi_f e_\mathbf{q} \nabla \Psi_i d_3 r. \tag{A.28}$$

It follows from Eq. (A.27) that the coefficient α, depending on the sign in front of the braces, can be negative, corresponding to absorption, or it can be positive, corresponding to amplification. The condition for positive α coincides with the conditions (8) for the presence of a negative temperature state. The maximum gain is obtained for $f_e(\varepsilon_i) = f_h(\varepsilon_f) = 0$, i.e., when all states in the valence band are filled, while in the conduction band they are free. It is reasonable that in this situation we should obtain the absorption coefficient calculated in [56]. The maximum gain occurs for a completely inverted distribution, when $f_e(\varepsilon_i) = f_h(\varepsilon_f) = 1$, and it is equal to the maximum absorption coefficient. In semiconductors, however, the absorption coefficient is so great in the case of indirect transitions that it exceeds by many orders of magnitude the coefficient of absorption by free carriers at the free-carrier concentrations required for the realization of an inverted distribution [39]. For an estimate of the gain, we introduce the quantity f [56] according to the equation

$$|P_{if}|^2 = \frac{1}{2}m\hbar\omega_{if}f_{if}, \tag{A.29}$$

where $f_{if} \approx 1$.

LITERATURE CITED

1. N. G. Basov, B. M. Vul, and Yu. M. Popov, Zh. Eksperim. i Teor. Fiz., 37 : 597 (1959).
2. N. G. Basov, O. N. Krokhin, and Yu. M. Popov, Zh. Eksperim. i Teor. Fiz., 39 : 1486 (1960).
3. N. G. Basov, O. N. Krokhin, and Yu. M. Popov, Zh. Eksperim. i Teor. Fiz., 40 : 1203 (1961).
4. N. G. Basov, O. N. Krokhin, and Yu. M. Popov, Advances in Quantum Electronics. Columbia University Press, New York (1961).
5. N. G. Basov, O. N. Krokhin, and Yu. M. Popov, Zh. Eksperim. i Teor. Fiz., 40 : 1879 (1961).

6. B. Lax, Quantum Electronics. Columbia University Press, New York (1960).
7. B. Lax, Advances in Quantum Electronics. Columbia University Press, New York (1961).
8. Yu. M. Popov, Fiz. Tverd. Tela, 5:1170 (1963).
9. N. Krömer, Proc. IRE, 47:397 (1959).
10. N. G. Basov, O. N. Krokhin, and Yu. M. Popov, Zh. Eksperim. i Teor. Fiz., 38:1001 (1960).
11. D. C. Matthis and M. J. Stevenson, Phys. Rev. Letters, 3:18 (1959).
12. P. Kaus, Phys. Rev. Letters, 3:20 (1959).
13. N. G. Basov, B. D. Osipov, and A. N. Khvoshchev, Zh. Eksperim. i Teor. Fiz., 40:1882 (1961).
14. B. D. Osipov and A. N. Khvoshchev, Zh. Eksperim. i Teor. Fiz., 43:1179 (1962).
15. N. G. Basov, O. N. Krokhin, L. M. Lisitsyn, E. P. Markin, and B. D. Osipov, Zh. Eksperim. i Teor. Fiz., 41:988 (1961).
16. B. M. Vul, A. P. Shotov, and V. S. Bagaev, Fiz. Tverd. Tela, 4:3676 (1963).
17. D. N. Nasledov, A. A. Rogachev, S. M. Ryvkin, and B. V. Tsarenkov, Fiz. Tverd. Tela, 4:1062 (1962).
18. R. N. Hall, G. E. Fenner, J. P. Kingsley, T. J. Soltus, and R. O. Carlson, Phys. Rev. Letters, 9:366 (1962).
19. M. I. Nathan, W. P. Kumke, G. Burns, F. H. Dill, and G. Lasher, Appl. Phys. Letters, 1:62 (1962).
20. V. S. Bagaev, N. G. Basov, B. M. Vul, B. D. Kopylovskii, O. N. Krokhin, E. P. Markin, Yu. M. Popov, A. N. Khvoshchev, and A. P. Shotov, Dokl. Akad. Nauk SSSR, 150:275 (1963).
21. V. A. Fabrikant, Tr. Vses. Elektrotekhn. Inst., 41:236, 254 (1950); Doctoral Dissertation, FIAN SSSR (1939).
22. E. M. Purcell and R. V. Pound, Phys. Rev., 81:279 (1951).
23. A. Einstein, Phys. Z. S., 18:121 (1917).
24. L. D. Landau and E. M. Lifshits, Quantum Mechanics. GITTL, Moscow (1948).
25. H. Frölich and B. V. Paranjape, Proc. Phys. Soc., B69:21 (1956).
26. L. V. Keldysh, Zh. Eksperim. i Teor. Fiz., 37:713 (1959).
27. V. A. Chuenkov, Fiz. Tverd. Tela, 2:200, 209 (1959).
28. J. Yamashita and M. Watanabe, Prog. Theoret. Phys., 12:443 (1954).
29. R. Stratton, Proc. Roy. Soc., A242:355 (1957).
30. Yu. M. Popov, Zh. Tverd. Fiz., 26:1634 (1956); 28:437 (1958).
31. L. V. Keldysh, Zh. Eksperim. i Teor. Fiz., 33:994 (1957).
32. B. M Vul, É. I. Zavaritskaya, and L. V. Keldysh, Dokl. Akad. Nauk SSSR, 135:1361 (1960).
33. B. A. Chuenkov, Proc. Internat. Conf. on Semiconductor Physics, p. 109. Prague (1960).
34. Yu. M. Popov, Zh. Eksperim. i Teor. Fiz., 35:505 (1958).
35. W. Kohn, Solid State Phys., 5:257 (1957).
36. M. Lax, Phys. Rev., 119:1502 (1960).
37. S. H. Koenig, Phys. Rev., 110:988 (1958).
38. W. C. Dunlap, Introduction to Semiconductors. Wiley, New York (1957).
39. W. M. Pumke, Phys. Rev., 127:1559 (1962).
40. N. G. Basov and A. M. Prokhorov, Usp. Fiz. Nauk, 57:485 (1955).
41. N. G. Basov, O. N. Krokhin, and Yu. M. Popov, Usp. Fiz. Nauk, 72:161 (1960).
42. L. D. Landau and E. M. Lifshits, Electrodynamics of Continuous Media. GITTL, Moscow (1957).
43. L. H. Hall, J. Bardeen, and F. J. Blatt, Phys. Rev., 95:559 (1954).
44. Herman, Phys. Rev., 88:1210 (1952).
45. J. R. Haynes, M. Lax, and W. F. Flood, Proc. Internat. Conf. on Semiconductor Physics. Prague (1960).
46. G. G. MacFarlane, I. P. McLean, J. E. Quarrington, and V. Roberts, Phys. Chem. Solids, 8:388 (1959).

47. J. R. Haynes, M. Lax, and W. F. Flood, Phys. Chem. Solids, 8:392 (1959).

48. S. Zwerdling, L. Roth, and M. Lax, Phys. Chem. Solids, 8:397 (1959).

49. H. Y. Fan, Usp. Fiz. Nauk, 64:733 (1958); 65:111 (1958).

50. V. S. Vavilov, Fiz. Tverd. Tela, 2:374 (1960).

51. H. Y. Fan, W. Spitzer, and R. J. Collins, Phys. Rev., 101:566 (1956).

52. H. J. G. Meyer, Phys. Rev., 112:298 (1958).

53. W. P. Dumke, Phys. Rev., 105:139 (1957).

54. G. G. MacFarlane and V. Roberts, Phys. Rev., 97:1714 (1955); 98:1865 (1955).

55. O. N. Krokhin, Master's dissertation. FIAN SSSR (1962).

56. J. Bardeen, F. J. Blatt, and L. H. Hall, Proc. Atlantic City Photoconductivity Conf.,
     New York (1954), (1956).

57. W. Heitler, Quantum Theory of Radiation. Oxford University Press (1954).

58. A. I. Akhiezer and V. B. Berestetskii, Quantum Electrodynamics. GITTL, Moscow (1956).

59. G. G. MacFarlane, I. P. McLean, J. E. Quarrington, and V. Roberts, Phys. Rev., 108:1377
     (1957).

60. O. N. Krokhin, Fiz. Tverd. Tela, 4:822 (1962).

61. E. Snitzer, Phys. Rev. Letters, 7:444 (1961).

62. N. G. Basov, L. M. Lisitsyn, and B. D. Osipov, Dokl. Akad. Nauk SSSR, 149:561 (1963).

63. F. J. McClung and R. W. Hellwarth, J. Appl. Phys., 33:828 (1962).

64. N. G. Basov, V. S. Zuev, and P. G. Kryukov, Zh. Eksperim. i Teor. Fiz., 43:353 (1962).

65. R. W. Hellwarth, Quantum Electronics. Proc. Third Internat. Congress, Paris (1963).

66. J. J. Laferski and P. Rappoport, Phys. Rev., 111:432 (1958).

67. A. F. Lammersfeld, Naturwissenschaften, 33:280 (1946).

68. Yu. M. Popov, Materials of the Seventh Conference on Luminescence, p. 281. Moscow (1958).

69. Yu. M. Popov, Fiz. Inst. Akad. Nauk SSSR, 23:67 (1963).

70. N. R. Whetten and A. B. Laporsky, Phys. Rev., 107:1521 (1957).

71. N. B. Gornyi, Zh. Eksperim. i Teor. Fiz., 35:281 (1958).

72. Dragokupil and Malkovskaya, Czech. J. Phys., 7:521 (1957).

73. V. S. Vavilov, Usp. Fiz. Nauk, 75:263 (1961).

74. W. Shockley, Czech. J. Phys., B11:81 (1961).

75. O. N. Krokhin and Yu. M. Popov, Zh. Eksperim. i Teor. Fiz., 38:1589 (1960).

76. B. I. Davydov, Zh. Eksperim. i Teor. Fiz., 7:1069 (1937).

77. W. Shockley, Electrons and Holes in Semiconductors. Van Nostrand, Princeton, New
     Jersey (1950).

78. F. Seitz, Phys. Rev., 73:549 (1948).

79. L. D. Landau and E. M. Lifshits, Statistical Physics. GITTL, Moscow (1951).

80. Yu. M. Popov, Opt. i Spektroskopiya, 7:697 (1959).

81. V. A. Chuenkov, Izv. Akad. Nauk SSSR, Ser. Fiz., 33:369 (1958).

82. N. Holonyak and S. E. Bevacqua, Appl. Phys. Letters, 1:82 (1963).

83. L. Bess, Phys. Rev., 105:1469 (1957).

84. V. L. Bonch-Bruevich and Yu. V. Gulyaev, Fiz. Tverd. Tela, 3:465 (1960).

85. V. L. Bonch-Bruevich, Coll.: Solid State Physics, Vol. 2, p. 182. Izd. Akad. Nauk SSSR,
     Moscow (1959).

REGENERATIVE LASER AMPLIFIERS

N. G. Basov, A. Z. Grasyuk, I. G. Zubarev, and L. V. Tevelev

INTRODUCTION

The present article concerns itself with an investigation of the properties of regenerative laser amplifiers and the methods for their practical application. The laser amplifier is a device in which coherent light is amplified. As in any quantum (maser) amplifier [1-3], amplification in the optical quantum (laser) amplifier is accomplished by virtue of induced emission in a medium with negative temperature.

It is required in some cases to amplify the coherent radiation emitted by an optical quantum generator (laser). For example, the laser amplifier is an essential component in any optical transmitting device when the emitting laser does not have adequate power. The dynamic range of the laser amplifier is extremely large, thus permitting the amplification of signals with very large power.

Laser amplifiers are also important as receivers of information-carrying coherent light signals, particularly in cases when the information is transmitted by phase or frequency modulation of the carrier. The high timewise coherence of the laser modes is such as to permit the use of electronic engineering techniques for the transmission of information and enhancement of the capacity of the communications channel. The transmission and reception of information in the optical range has a number of unique characteristics by comparison with the normal radio range. In particular, the quantum behavior of electromagnetic radiation begins to assert itself much earlier at low power levels in this case. The influence of the quantum structure of the radiation on the processes of information transmission and reception has been investigated in [4-7]. In one of these papers [6], the informational characteristics of various receiving devices are discussed, including quantum counters, coherent wave amplifiers, superheterodynes, etc.

Two operational modes are possible in the laser amplifier, namely the traveling-wave (without feedback) and the regenerative mode with positive feedback [7]. Open resonators are used to create positive feedback in the laser amplifier [8-11].

An experimental investigation of traveling-wave (ruby) laser amplifiers is described in [12]. The results of an experimental investigation of a multistage traveling-wave amplifier are also presented in [13, 14]. The laser amplifiers described in those papers are capable of amplification in one direction, i.e., they are unidirectional. The amplifier consists of several ruby crystals in series, decoupled by means of noninteracting elements operating on the principle of the Faraday effect. Such amplifiers are extremely useful in situations where it is necessary to amplify the energy (and hence the power) of the pulse delivered from a modulated-Q emitting laser [15, 16]. A theoretical investigation of the amplification of a light pulse in a medium with an inverted population is conducted in [17-19] using the equations of nonstationary transport.

As for any amplifier, the fundamental characteristics of the laser amplifier are the (power) gain, transmission (pass) band, and the sensitivity.

In the present article we give the fundamental theoretical relations required in order to assess the properties of regenerative laser amplifiers, the results of an experimental investigation of certain regenerative laser amplifier circuits, and their principal characteristics.

CHAPTER I

ELEMENTS OF AMPLIFIER THEORY

1. Gain and Pass Band of a Laser Amplifier

The feature that distinguishes the laser amplifier from its microwave counterpart (maser amplifier) is the fact that its dimensions are many times the wavelength. In determining the gain and pass band of the laser amplifier, therefore, it is advisable to use the methods employed for the investigation of radiation processes in free space [17-19]. Let there be a two-level system with a particle concentration n_2 at the upper level and n_1 at the lower level. The gain in decibels in the linear mode, according to [13], is equal to

$$K(\nu, \theta, \varphi, \mathbf{P}) = 10 \lg e\left[W(\theta, \varphi, \mathbf{P}) \frac{\lambda_0^2 l(\theta, \varphi)}{\varepsilon}(g_1 n_2 - g_2 n_1) g(\nu)\right], \tag{1}$$

where $W(\theta, \varphi, \mathbf{P})$ is the probability of spontaneous emission per unit time in unit solid angle for transition between the upper and lower levels in the direction θ, φ with polarization $\mathbf{P}$, $g(\nu)$ is the form factor of the spectral line, λ_0 is the wavelength of the incident radiation, l is the length of the active medium, ε is its dielectric constant, and g_1 and g_2 are the statistical weights of the lower and upper levels, respectively. For a line with Gaussian shape (ruby line R_1 at room temperature) we have

$$K = 10 \lg e\left[\frac{\lambda_0^2 l(\theta, \varphi)}{\varepsilon} W(\theta, \varphi, \mathbf{P})(g_1 n_2 - g_2 n_1) \frac{2}{\Delta \nu_l}\right] \sqrt{\frac{\ln 2}{\pi}} \exp\left[-\frac{(\nu - \nu_0)^2}{\Delta \nu_l^2} 4 \ln 2\right].$$

Here $\Delta \nu_l$ is the luminescence linewidth.

For an isotropic medium $g_1 W(\theta, \varphi, P) = \text{const} = 1/\tau_S$ and K [dB] assume the form

$$K = (10 \lg e) \frac{\lambda_0^2 l(\theta, \varphi)}{\pi \varepsilon \tau_s}\left(n_2 - \frac{n_1 g_2}{g_1}\right) \frac{2}{\Delta \nu_l} \sqrt{\frac{\ln 2}{\pi}} \exp\left[-\frac{(\nu - \nu_0)^2}{\Delta \nu_l} 4 \lg 2\right], \tag{1a}$$

where τ_S is the radiation lifetime of an active particle at the upper level 2, and ν_0 is the frequency corresponding to the center of the spectral line.

The pass band of a traveling-wave laser amplifier at the 3-dB level is given by the expression

$$\Delta \nu_k = \Delta \nu_l \sqrt{\frac{K_{\max}}{K - 3}}, \tag{1b}$$

where $K_{\max}$ is the gain (in decibels) at the center of the line ($\nu = \nu_0$). A regenerative laser amplifier with many modes in the resonator may be regarded as a traveling-wave amplifier for which the effective length L is larger than the corresponding quantity in a traveling-wave amplifier. The pass band of a regenerative amplifier becomes narrower with increasing gain G (dB), and the relationship of the latter to the amplifier band has the form

$$\Delta v_G = \Delta v_l \sqrt{\left(\lg \frac{G_{max}}{G_{max} - 3}\right) \Big/ \lg 2}. \tag{1c}$$

If the line has a Lorentzian shape

$$g(v) = \frac{1}{\pi} \frac{{}^1/_2 \Delta v_l}{(v - v_0)^2 + ({}^1/_2 \Delta v_l)^2}, \tag{1d}$$

we have for K, Δv_k, Δv_G, respectively,

$$\left. \begin{array}{l} K = 10(\lg e) \dfrac{\lambda_0^2 l(\theta, \varphi)}{8\pi e \tau_s} \left(n_2 - \dfrac{g_2}{g_1} n_1\right) \dfrac{1}{\pi} \dfrac{{}^1/_2 \Delta v_l}{(v - v_0)^2 + ({}^1/_2 \Delta v_l)^2}, \\[3mm] \Delta v_k = \Delta v_l \sqrt{\dfrac{3}{K_{max} - 3}}, \quad \Delta v_G = \Delta v_l \sqrt{\dfrac{3}{G_{max} - 3}}. \end{array} \right\} \tag{1e}$$

In cases when the signal power into the regenerative amplifier is so large that the inverted population of the medium cannot be assumed constant (nonlinear operation), the gain and pass band are best calculated by the methods used in [17-19].

2. Single-Mode Regenerative Laser Amplifier

The dimensions of regenerative amplifier resonators can be many times the wavelength. Consequently, in such amplifiers a great many modes are amplified simultaneously. Of course, the sensitivity of any amplifier with a plane-wave input, i.e., receiving a single oscillatory mode in space, is greater, the fewer the modes incident in the emission line. This means that the maximum sensitivity of the laser amplifier in this case is associated with the maximum restriction on the number of resonator modes whose frequencies fall within the limits of the spectral line. It is important, therefore, to consider some of the problems involved in the theoretical analysis of a laser amplifier with one oscillatory mode in the resonator.

The two-level single-mode regenerative laser amplifier was first investigated theoretically in [1, 2]. A large number of papers, one review of which is presented, for example, in [3], were subsequently devoted to the investigation of laser amplifiers. In [20-23] the process in such systems is analyzed by means of a closed system of equations having the form

$$\left. \begin{array}{l} \dfrac{d^2 E}{dt^2} + \dfrac{\omega_0}{Q} \dfrac{dE}{dt} + \omega_0^2 E = -4\pi n \dfrac{d^2 P}{dt^2} + \omega^2 F \cos \omega t, \\[3mm] \dfrac{d^2 P}{dt^2} + \dfrac{\omega_g}{Q_{l_2}} \dfrac{dP}{dt} + \omega_g^2 P = -\dfrac{2\mu^2_{ab} E\omega}{\hbar}(R + R_0), \\[3mm] \dfrac{dR}{dt} + \dfrac{1}{\tau_1} R = \dfrac{2E}{\hbar \omega_{ab}}\left(\dfrac{dP}{dt} + \dfrac{P}{\tau_2}\right), \end{array} \right\} \tag{2}$$

where $F \cos \omega t$ is the driving force, ω_0 is the natural frequency in the resonator, ω_{ab} is the frequency of the transition line, ω is the frequency of the external driving force, n is the particle density, μ_{ab} is the transition matrix element, and Q is the quality of the resonator,

$$\omega_g^2 = 1/\tau_2^2 + \omega_{ab}^2, \quad Q_{l_1} = \omega_g \tau_1/2, \quad Q_{l_2} = \omega_g \tau_2/2,$$

P is the polarization, R_n is the active-particle density, τ_1 and τ_2 are the relaxation times of R and P, respectively, E is the resonator field, F is the amplitude of the driving force, and $nR_0 = nR(t, 0)$, i.e., the number of active particles per unit volume at the initial instant of their interaction with the field (the pumping at the upper level is assumed to be constant). The various effects occurring in a regenerative laser amplifier can be analyzed rather thoroughly with the help of the system (2). In fact, as shown in [24], such a system is applicable at any time t, i.e., it describes any processes in the laser amplifier (and emitter), both stationary and nonstationary, it takes into account the phase of the oscillations and the nonlinearity of the system, and, in particular, it permits the superregenerative condition to be analyzed.

We introduce the following notation:

$$
\left.
\begin{aligned}
& X = \frac{\mu_{ab}\tau_2}{\hbar} E; \quad Y = \frac{P}{\mu_{ab} n R_0}; \quad u = \frac{R}{R_0}; \\
& \Delta_1 = \left[1 - \left(\frac{\omega_0}{\omega}\right)^2\right] Q; \quad \Delta_2 = \left[1 - \left(\frac{\omega_g}{\omega}\right)^2\right] Q; \quad t_1 = \frac{\omega}{2Q} t; \\
& \eta = \frac{4\pi \mu_{ab}^2 n R_0 \tau_2}{\hbar} Q; \quad w = \frac{\mu_{ab}\tau_2}{\hbar} F Q.
\end{aligned}
\right\}
\tag{3}
$$

From (1) we have

$$
\left.
\begin{aligned}
& \ddot{X} + \frac{\omega_0^2}{\omega^2} X = -\frac{1}{Q}\frac{\tau_1}{\tau_2}\eta\left(\frac{\omega_g}{\omega}\right)^2 Y + \frac{w}{Q}\cos\omega t - \frac{1}{Q_{l_2}}\frac{\omega_0}{\omega}\dot{X}, \\
& \ddot{Y} + \left(\frac{\omega_g}{\omega}\right)^2 Y = -\frac{1}{Q_{l_2}} X(u+1) - \frac{\omega_g}{\omega}\frac{1}{Q_{l_2}}\dot{Y}, \\
& \dot{u} = -\frac{\omega_g}{2Q_{l_1}\omega} + \frac{XY}{Q_{l_2}}.
\end{aligned}
\right\}
\tag{4}
$$

In accordance with the Van der Pol equation, we represent the solutions (2) in the form

$$
X = x(t)\cos(\omega t + \varphi); \quad Y = y(t)\cos(\omega t + \psi).
\tag{5}
$$

We obtain the following system of shortened equations for the amplitudes and phases:

$$
\left.
\begin{aligned}
& \dot{x} = \frac{\omega_0}{\omega} x + \frac{\tau_1}{\tau_2}\eta y \left(\frac{\omega_g}{\omega}\right)^2 \sin(\psi-\varphi) - w\sin\varphi, \\
& \dot{y} = \left(\frac{\omega_{ab}}{\omega}\right)^2 x\nu_2(u+1)\sin(\psi-\varphi) - \frac{\omega_g}{\omega}\nu_2 y, \\
& \dot{\varphi} = -\Delta_1 - \eta\frac{y}{x}\left(\frac{\omega_g}{\omega}\right)^2 \cos(\psi-\varphi) - \frac{w}{x}\cos\varphi, \\
& \dot{\psi} = -\Delta_2 + \nu_2\frac{x}{y}(u+1)\left(\frac{\omega_{ab}}{\omega}\right)^2 \cos(\psi-\varphi), \\
& \dot{u} = -\frac{\omega_g}{\omega}\nu_1 u - xy\nu_2\sin(\psi-\varphi),
\end{aligned}
\right\}
\tag{6}
$$

where $\nu_1 = Q/Q_{l_1}$, $\nu_2 = Q/Q_{l_2}$.

The equations describe the time dependence of the oscillation amplitudes and phases in the amplifier. The stationary solution, corresponding to the steady-state process, is readily found on the assumption that $\Delta_1 = \Delta_2 = 0$, letting $\dot{x} = \dot{y} = \dot{\varphi} = \dot{\psi} = \dot{u} = 0$ in the equations and expressing u + 1 and y in terms of x, whereupon we obtain a cubic equation in x:

$$
x_0^3 + w x_0^2 + (1 - \eta) x_0 + w = 0.
\tag{7}
$$

Consequently, the task of finding the stationary values of x_0, which are needed in order to analyze the laser amplifier, reduces to solution of Eq. (7). We bring it to the form

$$x_1{}^3 + 3px_1 + 2q = 0,$$

where

$$2q = \frac{2w^3 + 18w + 9\eta w}{27}; \quad 3p = \frac{3(1-\eta) - w^2}{3}.$$

(8)

The form of the solution of the cubic equation is determined by the quantities p, q.

Let us examine the solutions for values of p > 0 and q > 0, i.e., $\eta < (1 - w^2/3)$. Then $x_1 = -\frac{2}{9}\sqrt{3(1-\eta) - w^2} \sinh\zeta/3$, where $\sinh\zeta = q/|p|^{3/2}$. If we assume that $q/|p|^{3/2} \ll 1$, then $\sinh\zeta = \zeta = q/p^{3/2}$, and $\sinh\zeta/3 \simeq q/3p^{3/2}$. Then for x_1 we obtain

$$x_1 = -\frac{1}{9} \frac{2w^3 + 18w + 9\eta w}{3(1-\eta) - w^2},$$

(9)

whence

$$x_0 = \frac{1}{9} \frac{2w^3 + 18w + 9\eta w}{3(1-\eta) - w^2} - \frac{w}{3}.$$

It is clear from this expression that as $w \to 0$ we have $x_0 \to 0$, while for $\eta \to (1 - w^2/3)$ and $w \neq 0$ the amplitude grows, i.e., amplification takes place, the gain increasing as η approaches the self-excitation (auto-oscillation) threshold.

The power gain of this amplifier is determined by the expression

$$G = \left(\frac{x_0}{w}\right)^2 \simeq \frac{1}{(1-\eta)^2}.$$

Solution of the equation makes it possible to find the pass band of the amplifier and its dependence on the several parameters involved. It corresponds to the values of $\Delta = \Delta_1 = \Delta_2$ for which the following relation holds:

$$x_0(\Delta) = 0.7 \ x_0(0).$$

As mentioned earlier, the system (2) can be used to analyze the superregenerative condition of the amplifier as well, i.e., the case when the self-excitation condition $\eta > 1$ is satisfied. It is rather interesting in this connection to consider some of the effects associated with such a nonlinear condition. Of course, some of these effects are by no means inherent in ruby laser devices, inasmuch as the luminescence linewidth $\Delta\nu_l$ of the latter is considerably broader than the resonator band $\Delta\nu_r$. However, they are unquestionably inherent in laser devices utilizing materials for which $\Delta\nu_l$ is comparable (or nearly so) with $\Delta\nu_r$. The action of the external force on a two-level quantum system in a resonator has been investigated in [20-23] using the system (2) for the case when the self-excitation condition is fulfilled. It was established that "pulling" takes place under certain conditions in a laser; the oscillations in the laser occur synchronously with those of the driving force. Of course, this condition is not superregenerative in the complete sense of the word, since the amplitude of the oscillations depends nonlinearly on the amplitude of the driving force. However, it bears a certain interest when it is required to produce intense emission with a high time coherence by means of a laser emitter. As a matter of fact, the oscillation condition in a two-level quantum laser for which $\Delta\nu_l \leq \Delta\nu_r$ becomes unstable in strong overdriving. The theoretical possibility of such a condition was discovered in [25], wherein it was shown that the condition of instability is contained in the inequality

$$\eta = \alpha\nu > \frac{1+4\nu}{1-2\nu} \frac{1}{\nu}$$

(10)

or

$$\varphi(v) = 2\alpha v^3 - \alpha v^2 + 4v + 1 < 0. \tag{11}$$

Here $\nu = Q/Q_{l_1} = Q/Q_{l_2}$ (for $\tau_1 = \tau_2$).

In [24] the instability condition was derived for a laser with $\tau_1 \neq \tau_2$ in the form

$$\eta = \alpha v_1 > \frac{1 + 3v_2 + v_1}{1 - v_1 - v_2} \cdot \frac{1}{v_2}, \tag{12}$$

or in polynomial form

$$\Phi(v_1 v_2) = \alpha v_2^3 - \alpha v_2^2 (1 - v_1) + 3v_2 + v_1 + 1 < 0. \tag{13}$$

Negative values of Φ and φ, which define the region of instability, are possible when $\nu < 1/2$; $\nu_1 + \nu_2 < 1$. This means that the existence of such a condition is possible in lasers whose emission line is narrower than the resonator band at the working mode.

It was established in [24] that the amplitude of the oscillations in the laser emitter varies according to a quasi-harmonic law, i.e., an automodulation condition exists. This condition deteriorates the monochromaticity of the laser emission.

If, however, such an emitter is driven by an external signal, not only will synchronism set in under definite conditions, so will amplitude stabilization of the oscillations.† In other words, the external force removes the instability of the system (6), and stabilization occurs for an external signal whose amplitude is much smaller than that of stable stationary oscillations [23]. Consequently, the external signal "imposes" its own time characteristics on the laser, rendering the emission of the latter completely coherent. This condition may be regarded as a method of transforming a weak coherent signal into another coherent signal, more powerful and synchronized with the former, given a definite (albeit nonlinear) relation between those signals.

As shown in [23], amplitude stabilization sets in for an external force whose magnitude exceeds some threshold value, even if the frequency separation between that force and the auto-oscillations is zero, i.e., if in the system (6) $\Delta_1 = \Delta_2 = 0$. This threshold value of the external force is determined by the relation

$$w > \frac{2\alpha v^3 - \alpha v^2 + 4v + 1}{\frac{4}{3} v^2 \sqrt{\eta - 1} + (2\alpha v^3 - \alpha v^2 + 4v + 2) \frac{2}{3\sqrt{\eta}}}. \tag{14}$$

3. Sensitivity of a Regenerative Laser Amplifier

In order to realize the tremendous potentials afforded by coherent optics for the transmission of information, no small emphasis is to be placed on the sensitivity of the receivers and, in particular, the regenerative laser amplifier.

The sensitivity of any amplifier refers to the minimum signal delivered to its input so as to ensure a specified signal-to-noise ratio at the output. The limiting sensitivity of a quantum amplifier is determined by its internal noise. In the optical range $h\nu \gg kT$, i.e., the main noise source in the laser amplifier is spontaneous emission of a system of excited particles, which then determines the limiting sensitivity of every quantum amplifier. In the optical range the dimensions of the resonator are many times the wavelength. This means that many intrinsic

---

† Amplitude stabilization of a laser emitter operating in an unstable mode can also be accomplished by introducing negative feedback with respect to the Q of the resonator [23, 26].

modes will fall within the interval of the spectral line in general, their number being proportional to the probability of spontaneous emission and, hence, to the total spontaneous noise in the resonator. However, in different modes the spontaneous emission has different spatial and frequency characteristics.

If the reception of an information-bearing signal is accomplished over the entire space-frequency band, this is equivalent to the reception of information simultaneously in a large number of space- or frequency-separated channels [27]. The integral spontaneous noise of a regenerative laser amplifier in this case is distributed among different channels, hence it is possible in principle to have the case when each mode has its own channel. The reception of information via each such channel is realized in a single mode. The regenerative laser amplifier in this case is equivalent to a large number of laser amplifiers, each one operating in one mode, i.e., having the highest possible sensitivity and, consequently, transmitting power (capacity).

Thus, the presence of a large number of space- or frequency-separated modes in the resonator does not pose any conceptual difficulties in obtaining maximum information capacity on the part of the regenerative amplifier. All that is needed is a proper match between the number of space- and frequency-separated channels of the line and the number of modes in the regenerative amplifier.

The realization of this objective presents considerable engineering difficulties. It is necessary in practice to receive a signal whose space-frequency characteristics are given and do not coincide with those for the regenerative amplifier, for example, to receive and amplify radiation in the form of a plane wave, which represents one spatial mode. In this case, the regenerative amplifier must be designed so that spontaneous optical noise or other modes do not fall into the receiver stages connected to the output of the regenerative amplifier (for example, a photomultiplier). Otherwise the noise will be greater as the number of modes falling within the space-frequency band of the sensing device increases.

It is important to note that the dependence of the noise level on the number of modes N, in general, is different for different types of sensing devices. This problem has been investigated in [28], where it is shown, in particular, that the noise power for detectors is proportional to $\sqrt{N}$, while for a superheterodyne it is proportional to N.

Sensitivity of a Regenerative Laser Amplifier, Taking into Account the Discrete Structure of the Resonator Modes. The spontaneous noise power, which determines the sensitivity of a quantum amplifier, may be ascertained as follows in a single-mode resonator. The power of spontaneous emission P_n^{sp} is equal to

$$P_n^{sp} = N_2 h\nu W^{sp}, \tag{15}$$

where N_2 is the number of particles at the upper level and W^{sp} is the probability of spontaneous emission in the mode.

In correspondence with [1, 2], the power gain of a single-mode regenerative laser amplifier is equal to

$$G = \frac{\varkappa}{(1-\eta)^2}, \tag{16}$$

where $\varkappa < 1$ depends on the degree of coupling between the resonator and load. Hence $\eta = (G^{1/2}-1)/G^{1/2}$, where η is the self-excitation parameter, equal to

$$\eta = \frac{4\pi\mu^2(n_2-n_1)\nu}{\hbar\Delta\nu_l\Delta\nu_r} = \frac{4\pi\mu^2(n_2-n_1)}{\hbar\Delta\nu_l}Q. \tag{17}$$

Here μ is a matrix element of the dipole moment of the transition, $\Delta\nu_l$ is the linewidth, $\Delta\nu_r$ is the resonator bandwidth, and ν is the natural frequency of the resonator and coincides with the center of the line. According to [29], we have (provided $\Delta\nu_l \gg \Delta\nu_r$)

$$W^{sp} = \frac{4\pi^2}{V} \frac{\mu^2 \nu}{h\Delta\nu_l},$$ (18)

where V is the resonator volume.

From (15), taking (17) and (18) into account, we obtain the expression

$$P_n^{sp} = h\nu \frac{N_2}{N_2 - N_1} \frac{G^{1/2} - 1}{G^{1/2}} \Delta\nu_r,$$ (19)

where N_1 and N_2 are the numbers of particles at the upper and lower levels, respectively. The quantity P_n^{sp} attains a minimum value for constant $G \gg 1$ when $N_2 \gg N_1$. Then

$$P_n^{sp} = h\nu\Delta\nu_r.$$ (20)

In the case when there are N orthogonal modes regeneratable by the line the probability of spontaneous emission and, hence, the total noise in the resonator, increase N-fold (if we assume that all the modes have the same $\Delta\nu_r$), and the expression (20) assumes the form

$$P_n^{sp} = Nh\nu\Delta\nu_r.$$ (21)

In order to increase the sensitivity of a receiving device that includes a regenerative laser amplifier and receives one spatial mode, it is necessary to diminish the number of modes whose associated noise goes to the input of the succeeding receiver stages. This can be accomplished by reducing the number of modes regeneratable by the line, without substantially altering the increase in $\Delta\nu_r$. Consequently, the problem reduces to that of decreasing $N\Delta\nu_r$. If reception takes place in axial modes, as is normally done in practice, as this results in maximum amplification, it is necessary first of all to decrease $N_a\Delta\nu_r$, where N_a is the number of axial modes. The number of nonaxial modes can be reduced appreciably by spatial selection [14]. In order to reduce the number $N_a\Delta\nu_r$, it is necessary to have a resonator such that the natural frequencies of its axial modes will be separated by a sufficient distance in the vicinity of the line [30, 31]. The axial modes must have a sufficiently high Q in this case, i.e., they must have a small $\Delta\nu_r$. This is not feasible, for example, using the standard Fabry–Perot resonator with a simple reduction in length, because $\Delta\nu_r$ increases at the same time N decreases, such that the product $N_a\Delta\nu_r$ remains constant for constant r.

Thus, the frequency difference $\Delta\nu_a$ between adjacent axial modes in a resonator of the Fabry–Perot etalon type is $\Delta\nu_a = c/2L$, where L is the length of the resonator, c is the velocity of light. The number of axial modes ascribed to the line is

$$N_a = \frac{\Delta\nu_l 2L}{c}.$$

The pass band of a resonator operating in the given mode (neglecting diffraction losses) is equal to

$$\Delta\nu_r = \frac{(-\ln r)}{2\pi L/c}.$$

Here r is the reflection coefficient of the mirrors. Consequently,

$$N_a\Delta\nu_r = \frac{\Delta\nu_l(-\ln r)}{\pi},$$ (22)

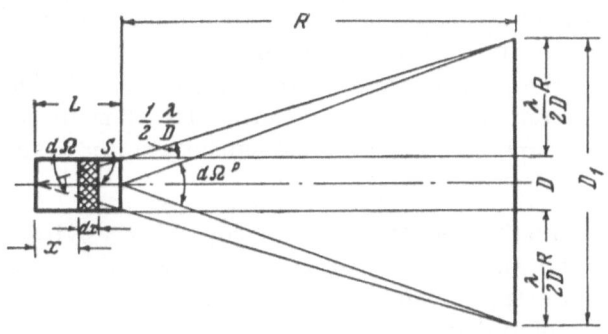

Fig. 1

i.e., for fixed r it does not depend on the length of the resonator. However, the spontaneous noise power in the resonator still depends on its length, since the latter (for a given power gain) is related to r. In fact, the gain of the regenerative laser amplifier is [14]

$$G = \frac{(1-r)^2 K}{1 + (Kr)^2 - 2Kr \cos \varphi} , \qquad (23)$$

where K is the single-pass gain and depends on the length of the resonators ($K = e^{\gamma L}$), r is the reflection coefficient of the mirror, and $\varphi = 2\pi l / \lambda$, l being the optical path in the resonator. When the match between the received radiation and the regenerative amplifier is a maximum ($\cos \varphi = 1$), we have for G

$$G = \frac{(1-r)^2 K}{(1 - Kr)^2} . \qquad (24)$$

Consequently, for constant G, the values of r and K are not independent. Expressing r in terms of K and G, and substituting the resultant expression into Eq. (21), we obtain the dependence of P_n^{sp} on K and, hence, on L (since $K = e^{\gamma L}$). From (24) we obtain for r

$$r = \frac{G^{1/2} K^{-1/2} - 1}{(GK)^{1/2} - 1}$$

and Eq. (21) acquires the form

$$P_n^{sp} = \frac{\Delta \nu_l}{\pi} \frac{(GK)^{1/2} - (G/K)^{1/2}}{(GK)^{1/2} - 1} \qquad (25)$$

or

$$P_n^{sp} = \frac{\Delta \nu_l}{\pi} G^{1/2} \frac{e^{1/2 \gamma L} - e^{-1/2 \gamma L}}{G^{1/2} e^{1/2 \gamma L} - 1} . \qquad (26)$$

For $\gamma L < 1$ we have for Eq. (26)

$$P_n = \frac{\Delta \nu_l}{\pi} G^{1/2} \frac{\gamma L}{G^{1/2} (1 + 1/2 \gamma L) - 1} . \qquad (27)$$

Hence, in order to obtain maximum sensitivity on the part of the regenerative laser amplifier, it is advisable to use resonators with small dimensions. The most promising in this respect are semiconductor materials [32, 33]. The gain per unit length in these materials is so high that the dimensions of the regenerative amplifier can be made several orders of magnitude smaller than for luminescent materials.

<u>Sensitivity of a Laser Amplifier without Regard for the Discrete</u>
<u>Structure of the Resonator Modes.</u> When the dimensions of the resonator are
much larger than the wavelength, the discrete structure of the different modes in the resonator
may be ignored. In this case, the regenerative laser amplifier may be thought of as a traveling-
wave laser amplifier whose active medium has a certain equivalent length, providing the necessary gain. The conditions for imparting maximum sensitivity to the laser amplifier may be determined as follows.

Let the isotropic active medium of the amplifier comprise a cylinder of length L, diam-
eter D, and cross section $S = \pi D^2 / 4$ (Fig. 1). Let D, $L \gg \lambda$, where λ is the wavelength of the
radiation, and suppose that reflections from the ends and lateral surface of the cylinder may be
neglected. Then the spectrum of eigenfrequencies of the volume occupied by the active medium
may be regarded as continuous and the methods of analyzing the interaction of electromagnetic
radiation with the substance in free space may be used. In this case (Fig. 1) the power of
spontaneous emission from a layer of thickness dx localized in the solid angle $d\Omega$ in the spectral
interval $(\nu, \nu + d\nu)$ is

$$dP_n^{sp} = \frac{h\nu n_2 S dx g(\nu) d\nu}{\tau_{sp}} \frac{d\Omega}{4\pi} ,\qquad(28)$$

where n_2 is the particle concentration at the uppermost of the two levels, $g(\nu)$ is the form factor
of the spectral line, and τ_{sp} is the radiation lifetime of a particle at the upper level.

The total spontaneous noise power P at the amplifier output is equal to the integral of the
expression (28) over x and ν:

$$P_n^{sp} = \int_{\nu_0 - \Delta\nu_l/2}^{\nu_0 + \Delta\nu_l/2} A \, d\nu \int_0^L e^{\gamma(\nu)(L-x)} dx,\qquad(29)$$

where

$$A = \frac{h\nu n_2 S g(\nu)}{\tau_{sp}} \frac{d\Omega}{4\pi} ,$$

ν_0 is the resonance frequency, $\Delta\nu_l$ is the luminescence linewidth, and γ is the amplification
factor of the active medium (in cm^{-1}).

According to [13], γ has the form

$$\gamma = \frac{\left(n_2 - \frac{n_1 g_2}{g_1}\right) \lambda_0^2}{8\pi\varepsilon\tau_{sp}} g(\nu).\qquad(30)$$

Here, λ_0 is the wavelength in vacuum, ε is the dielectric constant of the medium, g_1 and g_2 are
the degrees of degeneracy of the lower and upper levels, and n_1 is the particle concentration at
the lower level.

From (29), taking (30) into account, we obtain an expression for the spontaneous noise
power at the amplifier output:

$$P_{n,out}^{sp} \simeq 2h\nu \frac{S}{\lambda^2} d\Omega \left[K(\nu_0) - 1\right] \Delta\nu_K \frac{n_2}{n_2 - n_1 \frac{g_2}{g_1}} ,\qquad(31)$$

where $K(\nu_0) = \exp\left[\gamma(\nu_0)L\right]$ is the resonance gain, and $\Delta\nu_K$ is the amplifier pass band as determined by Eqs. (1b)-(1e).

Consequently, the signal-to-noise ratio at the amplifier output for the reception of a signal with power $P_{sig}^{(in)}$ and bandwidth $\Delta\nu_{sig}$ smaller than the amplifier pass band $\Delta\nu_K$ is

$$\theta = \frac{K(\nu_0) \, P_{sig}^{in}}{P_{n,out}^{(sp)}}, \qquad (32)$$

where $P_{n,out}^{(sp)}$ is determined by Eq. (31). In correspondence with the above definition of the amplifier sensitivity, $P_{sig\,min}^{(in)}$ is easily estimated by letting $\theta = 1$ in (32).

If a signal is received with a bandwidth exceeding the amplifier pass band ($\Delta\nu_{sig} > \Delta\nu_K$) and with a spectral density $p_\nu^{(in)}$, the power $P_{sig}^{(out)}$ of the amplified signal at the amplifier output is

$$P_{sig}^{(out)} = \int_{\nu_0-\Delta\nu_l/2}^{\nu_0+\Delta\nu_l/2} p_\nu^{(in)} K(\nu) \, d\nu \simeq p_\nu \Delta\nu_K K(\nu_0). \qquad (33)$$

The signal-to-noise ratio at the output in this case has the form

$$\theta = \frac{K(\nu_0) \, p_\nu^{(in)} \Delta\nu_K}{P_{n,out}^{(sp)}}, \qquad (34)$$

whence the minimum $p_\nu^{(in)}$ guaranteeing $\theta = 1$ is readily evaluated. For polarized radiation it is necessary in (31) to drop the coefficient 2, which corresponds to the two possible polarizations of the photon.

Let the arriving signal be a plane light wave comprising a single spatial mode. If the active medium of the amplifier is optically homogeneous, the amplified signal at the amplifier output will be localized in a solid angle $d\Omega = \lambda^2/S$, the size of which is determined by the diffraction of the plane wave at the exit (and entrance) aperture of the amplifier S (see Fig. 1). The aperture of the sensing device S_1 must be chosen so that the entire signal emanating from the amplifier output will impinge on the sensor. Then the latter will receive the spontaneous noise, which in accordance with (31) is proportional to its spatial bandwidth.

$$d\Omega^p = \frac{S_1}{R^2}.$$

As implied by Eq. (31) and Fig. 1, the minimum noise power at the amplifier output in this case occurs when

$$d\Omega^p \cong d\Omega = \frac{\lambda^2}{S}. \qquad (35)$$

The spontaneous noise power at the output of the laser amplifier is now

$$P_{n,min}^{out} = 2[K(\nu_0) - 1] \, h\nu \Delta\nu_K. \qquad (36)$$

As apparent from Fig. 1, the sensing device receives the entire amplified signal if $S_1 \geqslant \frac{\pi}{4}\left(\frac{\lambda}{D}R + D\right)^2$, where $S_1 = \pi D_1^2/4$ is the entrance aperture of the sensing device, i.e., the following relation must be fulfilled:

$$d\Omega^p \geqslant \frac{\pi\left(\frac{\lambda}{D}R + D\right)^2}{4R^2}. \qquad (37)$$

The relations (35) and (37) are fulfilled simultaneously if

$$R \gg \frac{D^2}{\lambda}.$$ (38)

Consequently, in order for the condition of maximum possible sensitivity to be realized in an optical amplifier, it is necessary, first, that the active medium be optically homogeneous and, second, that the spatial bandwidth of the sensing device be such that the output signal is sensed in the diffraction solid angle.

CHAPTER II

EXPERIMENTAL INVESTIGATION
OF THE REGENERATIVE LASER AMPLIFIER

1. Experimental Arrangement

We performed our investigations of the regenerative laser amplifier (RLA) using the arrangement shown in the block diagram of Fig. 2. The signal source was a ruby laser emitter (LE) with a modulated Q. The use of Q-modulation in the driving oscillator (single beam) greatly facilitated the measurements, in that the shape, duration, energy, and hence the power of the signal to be amplified were all known beforehand. Moreover, Q-modulation in the laser emitter made it possible to localize the amplified signal very precisely in time, thus facilitating the investigation of the time characteristic of the laser emitter. The emitter signal was attenuated by neutral light filters F, which had been calibrated previously. A collimator (theodolite) T was used to reduce the spread of the signal beam. The diameter of the amplified beam was controlled with an iris diaphragm D.

The laser amplifier operated on the circuits shown in Figs. 3-5 in two regimes, with and without Q-modulation of the resonator [15].

The receivers were photomultipliers; the signals were registered on a DÉSO-1 double-beam oscilloscope. In order to synchronize the firing of the laser emitter and amplifier lamps with the sweep of the oscilloscope, we used a timing device (system of electronic delay lines).

2. Gain

In a regenerative laser amplifier with a resonator of the Fabry–Perot type (Fig. 3), the input of the signal to be amplified is realized directly through a semitransparent mirror with reflection coefficient r_1, the output through a second mirror with reflection coefficient r_2. The power gain of such a single-pass regenerative laser amplifier is†

$$G = \frac{K(1-r_1)(1-r_2)}{1-K^2 r_1 r_2},$$ (39)

where K is the single-pass gain [14], as determined by the relation (1).

It is more suitable in practice to take energy from the resonator by means of auxiliary plates with a reflection coefficient r, which are situated inside the resonator (Fig. 4). The gain

---

† Equations (39)-(42) are derived by a procedure analogous to (23) with subsequent averaging over φ.

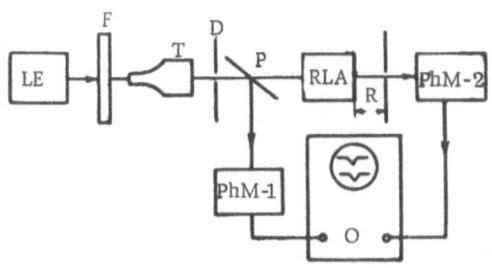

Fig. 2. Block diagram of the apparatus. LE) Laser emitter with modulated Q; F) attenuating filters; T) telescope system; D) diaphragm; P) semitransparent mirror; RLA) regenerative laser amplifier; R) distance from amplifier exit to registering device; PhM-2) photomultiplier for registration of the amplified signal; PhM-1) photomultiplier for registration of signal delivered to input of the laser amplifier; O) double-beam oscilloscope.

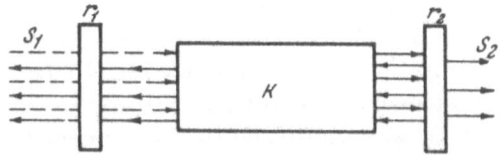

Fig. 3. Diagram of regenerative laser amplifier with resonator of the Fabry−Perot type. S_1) Input signal; r_1, r_2) mirror reflection coefficients; K) active medium; S_2) output signal.

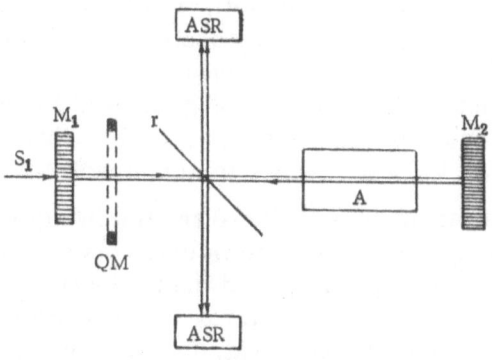

Fig. 4. Diagram of regenerative laser amplifier with regulated feedback. S_1) Input signal; M_1) entrance mirror; r) reflection coefficient of glass plate; ASR) amplified signal receivers; A) active medium; M_2) thick mirror.

after summation of both outputs of the laser emitter is

$$G = \frac{(1 - r_1) r}{1 - K^2 (1 - r) r_1} [1 + K^2 (1 - r)]. \quad (40)$$

The tuning of such systems is simple, but, as apparent from Fig. 5, which shows the dependence of G on the reflection coefficients of the mirrors of the single-pass regenerative laser amplifier, it is essential that the quantities K, r_1, and r_2 be chosen very accurately in order for G to have a given value in strong regeneration. Moreover, the approach to the oscillation threshold in such arrangements is highly critical, and to try and obtain a large gain with strong regeneration poses a rather difficult problem.

Its solution is greatly facilitated by the use of a polygonal resonator according to the scheme shown in Fig. 6.† The amplified beam in the resonator proceeds along a closed path (for example, via a triangle) in one direction. The input of the signal to be amplified is realized through a semitransparent mirror with reflection coefficient r. This type of directional arrangement for the regenerative laser amplifier possesses a number of advantages over the preceding types. Indeed, the gain of the regenerative amplifier in this case (for $r_1 = r_2 = 0$) is equal to

$$G = r + \frac{K (1 - r)^2}{1 - Kr}, \quad (41)$$

i.e., G > 1 for any K > 1 and 0 < r < 1. As experiments have shown, such a device is less critical to fluctuations in K and r, thus aiding considerably in the realization of a stable gain.

It is not uncommon in practice to meet situations in which r_1, $r_2 \sim r \neq 0$. For example, $r_1 = r_2 = 0.07$, unless a special optically transmissive coating is deposited on the ends of the ruby crystal. In this case, the resonator of the laser amplifier represents a complex system, the analysis of which results in the following relation between the power gain and the parameters K, r, r_1, r_2:

†Similar resonators have also been discussed in [34, 35] in application to laser emitters.

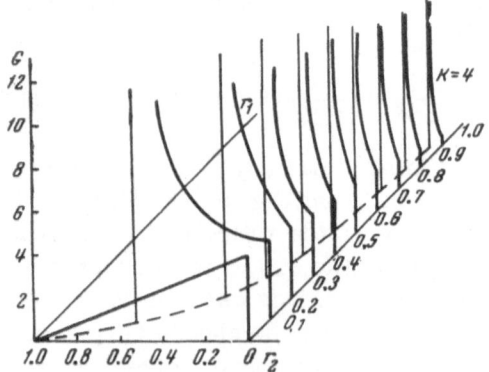

Fig. 5. Dependence of gain G on the reflection coefficients r_1 and r_2 of the resonator mirrors for the regenerative laser amplifier of Fig. 3.

$$G = G_0 + (1 - r)^2 \sum_{\substack{n=2m-1 \\ m=1}}^{\infty} , \; C_n^{2m-1} \frac{R^{m-1} G_{b_n}^n}{1 - G_{b_n} r} , \qquad (42)$$

where

$$G_{b_n} = \frac{K(1-r_1)(1-r_2)}{1 - K^2 r_1 r_2} , \quad G_0 = \frac{(1-r)^2 K}{1 - Kr} ,$$

$$R = \left[r_1 + \frac{(1-r_1)^2 K^2 r_2}{1 - K^2 r_1 r_2} \right] \left[r_2 + \frac{(1-r_2)^2 K^2 r_1}{1 - K^2 r_1 r_2} \right].$$

The dependence of G on K for r = 0.1, r_1 = r_2 = 0.07 is shown in Fig. 7. Also shown for comparison are curves corresponding to the case r_1 = r_2 = 0.

Furthermore, inasmuch as the input and output beams in such an amplifier are separated directionally, several such amplifiers may be joined in series without special decoupling arrangements between them. The block diagram for measuring the gain is equivalent to Fig. 2. As already noted, a laser with Q-modulation was used as the source. The amplified signal, on passing through the calibrated attenuation filter F, collimator T, and diaphragm D, was delivered to the regenerative laser amplifier and after amplification was sent on to the photomultiplier PhM-2, which acted as a registering device. Part of the signal was tapped off prior to its arrival at the amplifier by means of a plate P and sent to the photomultiplier PhM-1, which was used as a check on the amplitude of the input signal. The signal from the output of both photomultipliers was delivered to the double-beam oscilloscope DÉSO-1; the oscillograms were photographed. This procedure enabled us to measure both G and K (the single-pass gain), and to determine the stability of G and K, as well as their dependence on the various parameters involved.

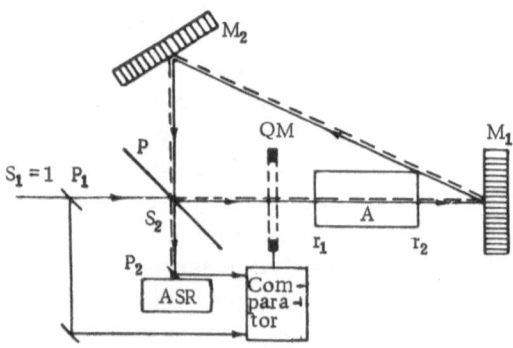

Fig. 6. Diagram of regenerative laser amplifier with triangular resonator. S_1) Input signal; P) semitransparent mirror with reflection coefficient r; ASR) signal receiver; QM) Q-modulation system; A) active medium; M_1, M_2) thick mirrors; r_1, r_2) reflection coefficients from ends of the crystal; P_1, P_2) plates designed to conduct a definite percentage of the input and output signals to the comparator.

These parameters included the following: the pumping energy and time (time interval between firing of the laser amplifier lamp and arrival of the signal to be amplified), and the feedback ratio, which was defined as the magnitude of the reflection coefficient r of the mirror. The results of the measurements (r = 0.1 for the laser amplifier arrangement shown in Fig. 6) are shown in Fig. 7, together with the theoretical curves; the agreement with the latter is seen to be acceptable.

A very promising technique for improving the parameters of regenerative laser amplifiers appears to be feedback modulation in the resonator (Q-modulation). Q-modulation makes it possible to create the negative temperature state beforehand. This eliminates fluctuations of the single-pass gain K (and hence the over-all gain G of the amplifier as well), since G does not

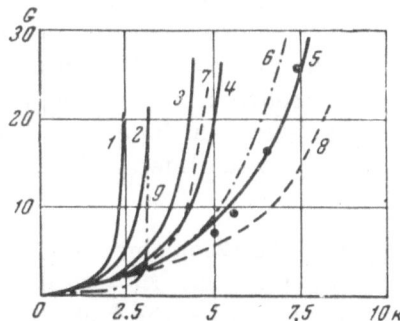

Fig. 7. Dependence of power gain G on the single-pass gain K for different values of positive feedback and various regenerative laser amplifier arrangements. Theoretical curves for the polygonal resonator arrangement (Fig. 6): 1) $r_1 = r_2 = 0$, $r = 0.4$; 2) $r_1 = r_2 = 0$, $r = 0.3$; 3) $r_1 = r_2 = 0$, $r = 0.2$; 4) $r_1 = r_2 = 0$, $r = 0.16$; 5) $r_1 = r_2 = 0$, $r = 0.1$; 6) $r_1 = r_2 = 0.07$, $r = 0.1$. Theoretical curves for the Fabry—Perot resonator arrangement: 7) according to Fig. 3 with $r_1 = r_2 = 0.2$; 8) for the same arrangement with $r_1 = r_2 = 0.1$; 9) according to Fig. 4 with $r_1 = r_2 = 0.1$. ••) Experimental points for polygonal resonator arrangement (Fig. 6) with $r = 0.1$, $r_1 = r_2 = 0.07$.

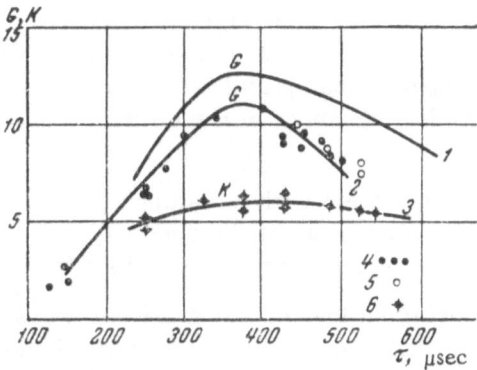

Fig. 8. Dependence of power gain G(K) and single-pass gain K on pumping time τ (μsec) of the amplifying crystal. 1) Theoretical dependence G(K); 2,3) experimental curves; 4) experimental points for G(K) without Q-modulation of amplifier resonator; 5) experimental points for G(K) with Q-modulation of amplifier resonator; 6) experimental points for K.

depend on the pumping power in this case, but is determined by the energy stored in the crystal at the moment of arrival of the signal. Moreover, the negative temperature state is preserved for a longer period of time in the absence of feedback than with the latter present, a fact that is plainly evident from physical considerations. In other words, the method of Q-modulation permits the negative temperature state to "persist" until arrival of the signal to be amplified, provided, of course, this signal does not exceed the radiation lifetime of the active particles in free space.

The results of the pertinent measurements are shown in Fig. 8, from which it is apparent that the gain for the amplifier without Q-modulation decays with time more rapidly than the equivalent gain for the laser with Q-modulation.

For stabilization of the laser amplifier gain it is permissible also to use negative feedback with respect to the quality (Q) of the resonator (see Fig. 6), the influence of which on the operation of a laser has been investigated in [23, 26].

A certain part of the input signal (for example, S_1/m) is tapped off by the plate P_1 to the comparator (input-output comparison network). At the same time, the plate P_2 is used to tap off an appropriate part of the output signal (for example, S_2/mG, where G is the specified gain). If $S_2/G \neq S_1$, an error signal appears at the output of the comparator; the error signal is delivered in appropriate phase to the element QM with a regulated transmissivity. By regulating the reflection coefficient of P_2 it is possible to control the value of the stabilized over-all gain G.

3. Measurement of the Sensitivity

In our case, the dimensions of the resonator are several orders of magnitude times the wavelength. This means that the discrete structure of the resonator modes may be neglected and the relations (31) and (32) used to estimate the sensitivity of the amplifier, letting $\theta = 1$. As apparent from (31) and (32), to achieve maximum sensitivity, it is necessary to limit as much as possible $d\Omega^p$, the spatial bandwidth of the registering device. As shown

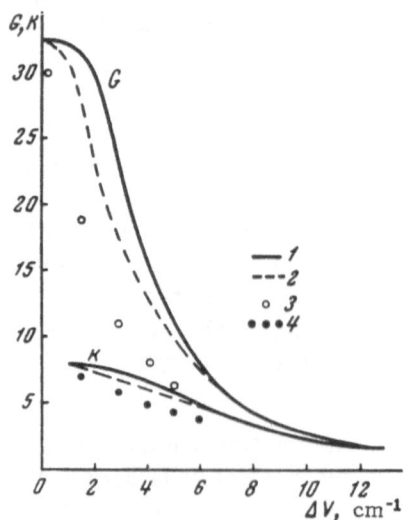

Fig. 9. Dependence of the gains G and K on the frequency. Theoretical curves: 1) Gaussian line configuration; 2) Lorentzian line configuration. Experimental points: 3) measurement of G; 4) measurement of K.

in §3 of Chapter I, the minimum value of $d\Omega^P$ cannot be less than λ^2/S, because the signal to be amplified is concentrated in this angle due to diffraction at the exit aperture of the laser amplifier. Moreover, in this case, the parameters of the registering device which determine its spatial bandwidth must be chosen with allowance for the relation (38).

The limiting sensitivity of the laser amplifier, therefore, for $d\Omega^P \simeq \lambda^2/S$, $n_2 \gg n_1(g_2/g_1)$, $K(\nu_0) \gg 1$, is $P^{(in)}_{sig\,min} = h\nu\Delta\nu_K$. Here we have omitted the factor 2 in (31), since the emission from ruby is polarized. For a ruby laser amplifier $h\nu = 3\cdot10^{-19}$ J, $\Delta\nu_K \simeq 3\cdot10^{11}$ cps, whence $P^{(in)}_{sig\,min} = 10^{-7}$ W.

The sensitivity of the laser amplifier was measured using the scheme shown in Fig. 2. The driving emitter was a laser with modulated Q (one light pulse), which made it possible to determine the power of the input signal to the regenerative laser amplifier very precisely. The light pulse from the laser source was transmitted through the telescope system T, at the output of which was obtained a beam with an angle of spread less than the diffraction angle at the input and output aperture of the amplifier.

One sweep of the oscilloscope was timed to receive a signal from the photomultiplier PhM-1 (see Fig. 2), at which the glass plate P was used to tap part of the signal received at the input of the laser amplifier. This photomultiplier was used to monitor the power of the driving emitter. The second sweep of the oscilloscope was used to receive a signal from the PhM-2, which acted as a registering device. Both the amplified light beam from the driving emitter and the spontaneous noise of the amplifier were delivered to this second photomultiplier. The emitter beam was attenuated beforehand by calibrated light filters until the signals from the driving emitter and the spontaneous noise of the amplifier had become equal in magnitude on the oscilloscope. The power of the light signal received at the input of the optical amplifier, being equal in this case to its sensitivity, was determined from the known power of the emitter and the total signal attenuation factor.

The diameter of the receiving diaphragm (PhM-2) was $D_1 = 2$ cm with a distance $R = 10$ m and $S = 0.3$ cm. Consequently, $d\Omega^P = 4\cdot10^{-6}$ ster and the expected value of the sensitivity $P^{(in)}_{sig\,min} = h\nu\,(S/\lambda^2)d\Omega^P = 2.4\cdot10^{-5}$ W. The measured value of the sensitivity, $3-5\cdot10^{-5}$ W, agrees satisfactorily with the theoretical value. The value obtained for the sensitivity is worse than the limiting value by more than two orders of magnitude. This is because the spatial bandwidth of the registering device was $4\cdot10^{-6}$ ster, i.e., two orders of magnitude greater than the minimum possible $d\Omega^P = \lambda^2/S$. It was impossible to reduce $d\Omega^P$ any further, insofar as the optical inhomogeneity of the amplifier crystal caused the amplified light to be localized approximately within an angle of $4\cdot10^{-6}$ ster. In order to achieve the limiting value of the sensitivity [37] as determined according to (20), it would be essential to use optically homogeneous crystals.

It must not be overlooked that the limiting sensitivity is observed for a band of about $3\cdot10^{11}$ cps. If the received signal has a narrower band, it is necessary to match the band of the receiving laser amplifier with the band of the signal, as is done in electronic engineering. The limiting spectral sensitivity of the laser amplifier is high, $3\cdot10^{-19}$ W/cps. Consequently, optical quantum amplifiers are expected to find applications as receivers in communication lines.

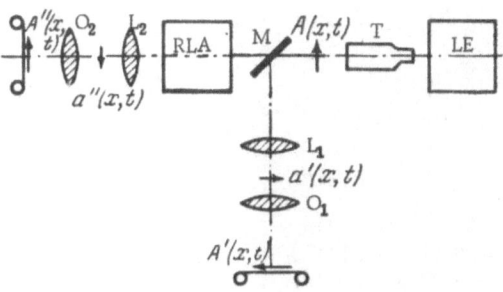

Fig. 10. Block diagram for the magnification of an image by means of a regenerative laser amplifier. LE) Laser emitter (with modulated Q); T) telescope system; A(x, t) − semitransparent object; P) plate used to tap part of the input signal; L_1) lens; $a'(x, t)$ − intermediate image; A'(x, t) − reference image (image received at the laser amplifier input); O_1) camera objective; L_2) lens; $a''(x, t)$ − intermediate image; O_2) camera objective; A"(x, t) − magnified image.

It is imperative once again to emphasize the opportunities offered by the development of laser amplifiers using semiconductor materials. Their small dimensions permit the limiting sensitivity to be attained very easily. In fact, as Eq. (35) implies, for D ~ $\sqrt{S} \simeq 0.4$ mm, the limiting spatial bandwidth of the registering device (sensor) amounts to $\lambda^2/S = (7 \cdot 10^{-4}/0.04)^2 \simeq 3 \cdot 10^{-4}$ ster, which is entirely practicable.

4. Measurement of the Regenerative Laser Amplifier Pass Band

The pass band of the regenerative laser amplifier was measured by a method similar to that used in electronics; the frequency of the emitter was varied and the frequency dependence of the gain measured.

In order to vary the emission frequency of the laser, we made use of the temperature shift of the ruby line [36] with cooling of the crystal.

The measured temperature dependence of the gains K(T), G(T) was used to determine the pass band of the amplifier:

$$\Delta\nu = 2\,(\nu_0 - \nu)\,|_{K = 1/2\,K_0},$$

where ν_0 is the frequency corresponding to the middle of the central part of the line at room temperature, ν is the emission frequency determined by the temperature of the cooled emitter crystal, K_0 is the gain with the amplifier and emitter crystals at the same (room) temperature.

The results of measuring the dependences K($\Delta\nu$), G($\Delta\nu$), where $\Delta\nu = \nu - \nu_0$, are shown in Fig. 9.

The theoretical curves were calculated according to Eq. (1) for lines with Gaussian and Lorentzian form factors.

As evident from Fig. 9, the frequency characteristic of the regenerative laser amplifier departs from a Gaussian configuration, more nearly approaching the Lorentzian.

5. Image Magnification with the Regenerative Laser Amplifier

In a number of instances there is no need for the laser amplifier to operate in just one mode. As a matter of fact, if sufficiently many modes with wave vectors in different directions fall within the linewidth, the laser amplifier can amplify light traveling in different directions. This relates to the fact that plane light waves traveling in different directions and characterized by different wave-vector directions fall within the realm of different modes of the amplifier. The total number of these modes determines the spatial bandwidth of the laser amplifier, i.e., the solid angle relative to some chosen direction (usually the direction characterized by maximum gain), from which the amplifier is capable of receiving and amplifying signals with a prescribed gain.

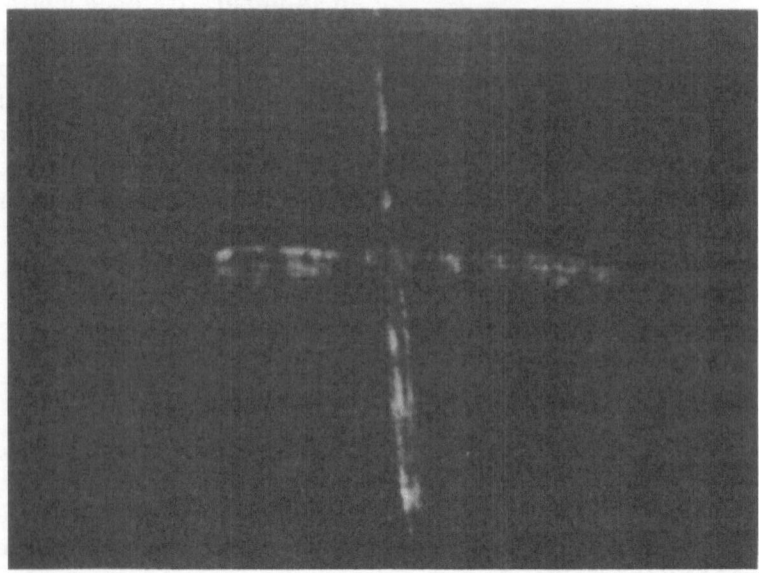

Fig. 11. Reference image A'(x, t) [image of A(x, t) received at
input to the regenerative laser amplifier].

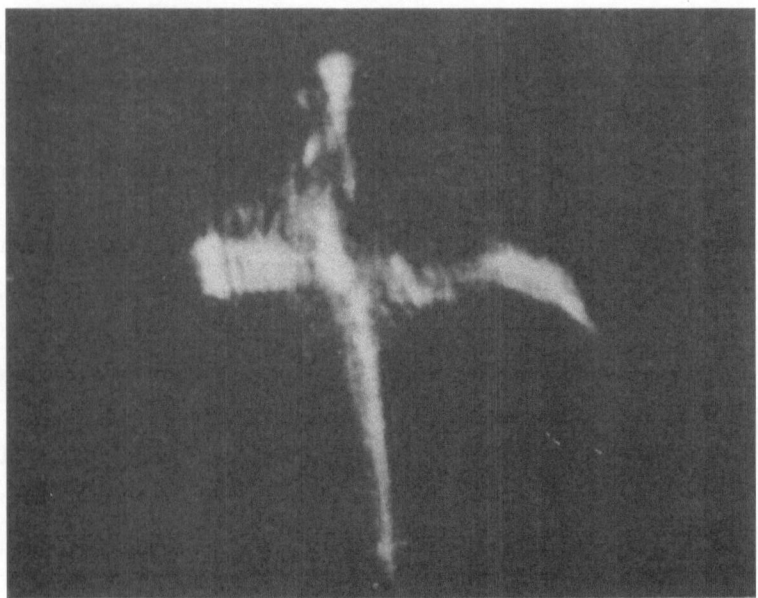

Fig. 12. Magnified image.

This property of the laser amplifier makes it possible to amplify (magnify) the image of
some semitransparent object illuminated with a laser beam [13]. The transmission and reception of the image are equivalent to the transmission of information simultaneously over a large
number of spatially separated channels [13, 38]. This method of transmitting and receiving information makes it possible to build a multichannel communication line, the capacity of which is
considerably larger than in the case of information transmission via a single channel, given the
same transmitter power rating.

Image magnification is possible either by means of a traveling-wave laser amplifier [13] or with a regenerative laser amplifier. This particular method of transmitting and receiving information has already made it possible to explore the feasibility of matching the transmitted information with the space-frequency band of a laser amplifier or regenerative laser amplifier, as mentioned in the discussion above. The results of some preliminary experiments on the magnification of an image by means of regenerative laser amplifiers according to the scheme of Fig. 6 are presented here. The measurement procedure was similar to that used in [13], where image magnification was accomplished by means of a traveling-wave laser amplifier. A block diagram showing the setup for the measurements is given in Fig. 10. The collimated beam from the laser emitter (LE) was used to illuminate the object $A(x, t)$. Half of the energy was taken off into the photographic attachment O_1 by means of the semitransparent mirror M. An auxiliary lens L_1 was placed in front of the camera objective, producing an intermediate image $a'(x, t)$. The camera objective was situated such that $a'(x, t)$ would be located between its focus and twice the focal length. This permitted the image of $A(x, t)$ to be magnified and its structure to be investigated.

The receiving system with its camera O_2 was identical to the one just described. The results of the preliminary experiments are shown in Figs. 11 and 12. For R = 1 m, the thickness of the cross d = 0.3 mm. As seen in Fig. 12, the image is distorted by the inhomogeneity of the regenerative laser amplifier crystal.

LITERATURE CITED

1. N. G. Basov and A. M. Prokhorov, Usp. Fiz. Nauk, 57(3) : 485 (1955).
2. N. G. Basov, Doctoral Dissertation, FIAN AN SSSR (1956).
3. G. M. Zverev, N. V. Karlov, L. A. Kornienko, A. A. Manenkov, and A. M. Prokhorov, Usp. Fiz. Nauk, 77(1) : 61 (1962).
4. T. E. Stern, IRE Trans. Inform. Theory, IT-6 : 435 (September, 1960).
5. D. S. Lebedev and L. B. Levitin, Dokl. Akad. Nauk SSSR, 149(6) : 1294 (1963).
6. I. P. Gordon, Proc. IRE, 50(9) : 1898 (1962).
7. N. G. Basov, A. Z. Grasyuk, and I. G. Zubarev, Dokl. Akad. Nauk SSSR, 157 : 1084 (1964).
8. A. M. Prokhorov, Zh. Eksperim. i Teor. Fiz., No. 6 (1958).
9. A. Schawlow and C. H. Townes, Phys. Rev., 112 : 1940 (1958).
10. A. G. Fox and T. Li, Bell System Tech. J., 40(2) : 453 (1961).
11. L. A. Vainshtein, Zh. Eksperim. i Teor. Fiz., 45(3) : 684 (1963).
12. P. P. Kisliuk and W. A. Boyle, Proc. IRE, 49 : 1635 (1961).
13. J. E. Geusic and H. E. D. Scovill, Bell System Tech. J., 41 : 1371 (1962).
14. J. E. Geusic and H. E. D. Scovill, Quantum Electronics. Proc. Third Internat. Congress, Paris (1964).
15. N. G. Basov, V. S. Zuev, and P. G. Kryukov, Zh. Eksperim. i Teor. Fiz., 43 : 353 (1962).
16. R. W. Hellworth, Advances in Quantum Electronics. Columbia University Press, New York (1961).
17. L. M. Frantz and J. S. Noduik, J. Appl. Phys., 34(8) : 2347 (1963).
18. M. L. Ter-Mikaélyan and A. L. Mikaélyan, Dokl. Akad. Nauk SSSR, 155(6) : 1298 (1964).
19. N. G. Basov and V. S. Letokhov, Opt. i Spektroskopiya, 18 : 1042 (1965).
20. A. N. Oraevskii, Radiotekhn. i Elektron., 4(4) : 718 (1959).
21. R. V. Khokhlov, Radiotekhn. i Elektron., 3(4) (1958).
22. N. K. Maneshin and R. V. Khokhlov, Vestn. Mosk. Univ., Ser. Mat., Mekhan., Fiz., Khim., No. 2, p. 109 (1958).
23. A. S. Agabekyan, A. Z. Grasyuk, I. G. Zubarev, A. N. Oraevskii, and V. I. Svergun, Radiotekhn. i Elektron., 9(12) : 2156 (1964).

24.   A. Z. Grasyuk and A. N. Oraevskii, Radiotekhn. i Elektron., 9(3) : 524 (1964).
25.   A. S. Gurtovnik, Izv. Vysshikh. Uchebn. Zavedenii Radiofiz., Vol. 1, No. 5-6 (1958).
26.   F. R. Marshall and D. Roberts, Proc. IRE, 50(10) : 2108 (1962).
27.   N. G. Basov, A. Z. Grasyuk, and A. N. Oraevskii, Radiotekhn. i Elektron., Vol. 9, No. 5 (1964).
28.   N. V. Karlov and A. M Prokhorov, Radiotekhn. i Elektron., 9(12) : 2088 (1964).
29.   F. V. Bunkin and A. N. Oraevskii, Izv. Vysshikh. Uchebn. Zavedenii Radiofiz., 2(2) : 181 (1959).
30.   S. A. Collins and G. R. White, Quantum Electronics. Proc. Third Internat. Congress, Paris (1963).
31.   H. Magner and H. Rote, Phys. Letters, 7(5) : 330 (1963).
32.   N. G. Basov, O. N. Krokhin, and Yu. M. Popov, Usp. Fiz. Nauk, 72 : 169 (1960).
33.   N. G. Basov, Third International Conference on Quantum Electronics. Paris-New York (1964).
34.   L. A. Kulevskii, P. P. Pashinin, and A. M. Prokhorov, Third International Conference on Quantum Electronics. Vol. II, Paris-New York (1964).
35.   A. Schawlow, Sci. Am., 209(1) : 34 (1963).
36.   D. E. McCumber and M. D. Sturge, J. Appl. Phys., 34 : 1682 (1963).
37.   N. G. Basov, A. Z. Grasyuk, and I. G. Zubarev, Zh. Prikl. Spectroskopiya, 3(1) : 26 (1965).
38.   J. E. Rosental, J. Appl. Opt., 1(2) : 169 (1962).

POWER PULSATION MODES OF LASERS

A. N. Oraevskii and A. V. Uspenskii

INTRODUCTION

The power pulsing of laser emission is a problem widely discussed in the literature [1-55]. It was first observed in masers (see, e.g., [25]), but the problem aroused particular interest when the same mode of operation was detected in optical masers, or lasers [1, 2].

A great many theoretical hypotheses have been advanced to date to account for the pulsing modes, or spiking, but not one of them has given a satisfactory description of all the experimental facts.

We begin with a brief summary of the experimental data. Three different oscillatory modes are observed: chaotic pulsing (spikes succeeding one another at different time intervals with dissimilar heights and shapes); periodic pulsing; and, finally, quasi-stationary states. All three states are observed in the ruby laser and in lasers using glass or a crystal activated with neodymium. In other solid-state lasers, for example [26], only the third mode of operation is observed. We will investigate these modes in further detail.

Chaotic Pulsing Condition. This condition is observed in a ruby laser with plane mirrors [1, 2, 10], in ruby lasers with spherical mirrors [16], in a neodymium-activated crystal or glass laser with plane mirrors [15]. The following data have been obtained for ruby. The average repetition frequency of the spikes is about 1 Mc;† this frequency increases with increasing pumping power and decreasing temperature [3, 9]. Moreover, with a reduction in temperature the regularity of the spikes increases, and a constant component makes its appearance ([7] and, clearly, [8]). During each spike several modes are generated simultaneously [4, 9, 12]; besides, individual parts of the end face, sometimes overlapping ones, generate different frequencies [5]. The chaotic pulsing condition has not been studied in detail in neodymium lasers. It is noted in [15] that such a condition is typical of samples with an emission spectrum in the form of narrow lines and a small angular distribution on the part of the emitted radiation.

Periodic Pulsing Condition. This condition obtains in a ruby laser under the following circumstances: in a laser with confocal mirrors at room temperature [10, 24] and at the temperature of dry ice [10]; in a laser with spherical mirrors [16], for which a very narrow emission spectrum has been noted [17] (0.1 cm^{-1}, i.e., about one order of magnitude smaller than for ruby with plane mirrors). It was shown in [24] that the frequency of the periodic pulsa-

---

† On varying the separation between mirrors (in a resonator with spherical mirrors) and the pumping power, the frequency of the spikes may suffer a severalfold variation; 1 Mc is the spike frequency under the standard operating conditions for a ruby laser.

tions increases with the pumping power. For a neodymium-activated crystal or glass laser with plane mirrors, the periodic pulsing condition has only been observed for certain samples [15], a typical characteristic of which has been a small angular divergence of the emitted radiation (less than 6°), whereas the quasi-periodic condition has been observed for samples with a divergence of 8° or more. The emission spectrum is made up of continuous bands. In all the samples for which the periodic condition has been observed, with a small pumping power damped pulsations have appeared, and only with a sufficient increase in the pumping power has the regular pulsing condition occurred. If the Q of the resonator is diminished (by placing a diffuser between the mirror and crystal), the regular pulsing condition vanishes. The regular pulsing condition has also been successively observed when a plane reflector was placed at a short distance from a laser with plane mirrors [15]. This condition has also been obtained in long glass filaments activated with neodymium using high pumping levels [15]. It is interesting to note the existence of samples in which it has generally not been possible to create the regular pulsing condition [15].

Quasi-Stationary Condition.   This condition has been observed in toroidal ruby lasers [6, 23], in ruby lasers with plane mirrors [14], in ruby lasers with plane mirrors at room temperature and high pumping power† [10], in ruby in a confocal resonator cavity at liquid nitrogen temperature and high illuminations [10], in a neodymium-activated crystal or glass laser with plane mirrors [15], and in ruby lasers with a spherical resonator [16]. The condition has been observed also in [13], where a system comprising a plane and a concave mirror with focusing lens between them was used. Typical of this condition is the presence of a short segment of damped pulsations. Their frequency, both in a ruby and in a neodymium-activated crystal or glass laser, behaves the same as the spiking frequency in the chaotic pulsation condition. The decay time diminishes with cooling [13] and increased pumping [6, 15].

An interesting result is obtained in the case of a neodymium-activated crystal or glass laser [15]. It has been shown that if the emission spectrum consists of several bands, the emission pattern acts as through active particles were burned out in the band (although after several, rather than just one spike).

The circumstances under which one of the above modes would go over to another have been given little attention. The results of [16] are significant from this standpoint, where all three modes were observed using the same apparatus and the same crystals, depending on the position of the mirrors and the pumping level.

In the case when the distance between the mirrors was on the order of twice the radius of curvature of the mirrors, regular pulsations were observed as a rule at low pumping powers. An increase in the pumping power caused a reduction in the percentage modulation and, clearly, changeover to the quasi-stationary condition, whereas an increase in the distance between mirrors led to a gradual transition from the periodic to the chaotic pulsation condition. Moreover, two of the modes were observed at once in a series of oscillograms (the chaotic pulsation mode first, then the periodic mode), the one condition being substituted by the other very abruptly.

It is important in summarizing the experimental results to realize the complexity of the emission pattern of lasers; it is not always clear from experimental investigations just which parameter variation has resulted in the change of mode. The modes in a ruby laser and a neodymium-activated crystal or glass laser are very similar, the same conditions being observed in resonators of diverse configuration. This fosters the notion that neither the shape of the resonator nor the nature of the spectral linewidth of the substance‡ is decisive with regard to

---

† It is possible that a small-amplitude pulsing condition was observed here.

‡ For ruby the line broadens uniformly; for neodymium crystals and glasses inhomogeneous broadening prevails.

the pulsing mode. The pulsation pattern is further complicated by the fact that certain spikes in the emitted radiation from a ruby laser are modulated in turn [18-22] with frequencies ranging from a few megacycles to several hundred megacycles. Some photographs presented in [22] show that different parts of the end face can be modulated with different frequencies, and even if with the same frequency, then not in the same phase. Several frequencies can be observed in a single photograph. The frequency of the modulated spikes depends on the temperature and pumping level [21]. There is no relationship between the modulation and the optical quality of the crystal [18]. The percentage modulation does not depend on the pumping [20]. Yet double modulation is not observed in a neodymium-activated crystal or glass laser [19].

We now analyze the basic theoretical concepts relating to the theory of the pulsations.

1. The pulsations simply constitute a stabilization or settling process in the laser (such an inference is drawn, for example, from an investigation of the rate equations). If the settling process is a sufficiently long-time one and we use a pulsed lamp for pumping, the emission of the laser will have a pulsating character throughout the duration of the pulse. However, for ruby at least this postulate is not necessarily valid, since a ruby laser also pulsates in continuous operation [27].

2. A model of holes burning in the line has been proposed [4, 28]. The theory is essentially the following. The self-excitation condition is fulfilled within the limits of the spectral line at some frequency ν_1, and the field amplitude begins to grow (ascending portion of the spike). If relaxation within the limits of the line proceeds more slowly than the de-excitation of the active particles at the frequency ν_1, then right away at this frequency a "burning" of the active particles takes place, a diminution of the field is observed (descending portion of the spike), and, finally, oscillation ceases. The next spike occurs in another part of the line, and so on. It follows from the theory, first, that the spectral line must be inhomogeneously broadened (this is certainly not so for ruby) and, second, the observed spikes, in general, must be asymmetric (the buildup and decay must behave differently), a fact that has not been corroborated by experiment.

3. Some authors have proposed a model of spatial "burning." The beam is regarded as the "burning" of active particles in one mode, but inasmuch as different modes have different spatial distributions, spatial burning occurs.

4. The opinion has been expressed that the pulsing is attributable to nonlinear interactions between the modes.

5. An attempt has been made to relate the pulsing to noise [31].

6. Finally, many authors have advanced the notion that the pulsing is related to nonlinear effects of the field interacting with the material in the resonator, and the theory of the pulsing can be simplified by treating two- and three-level systems [29, 30].

All of the above theoretical hypotheses refer to amplitude modulation with a frequency on the order of 1 Mc. With regard to the second, more rapid, modulation observed in ruby, so far no hypotheses have been offered other than the notion of beats between the modes.

The investigation of two- and three-level systems in a resonator with one mode has been given the most attention.

In the present article we clarify the nature of the nonstationary processes in a two-level system within the framework of a single mode, along with the conditions under which they occur. We will use the τ_1, τ_2 model for the active substance.

1. Equations for a Two-Level Laser (τ_1, τ_2 Model)

Consider a two-level laser with $w_2 > w_1$. Let a constant or nearly constant flux of active particles I be admitted at the upper level. We assume that one mode is excited in the resonator and neglect spatial effects. Then the equations describing the oscillatory processes in such a laser have the form

$$\left.\begin{aligned}
\ddot{\mathscr{E}} + \frac{\omega_r}{2Q}\dot{\mathscr{E}} + \omega_r^2\mathscr{E} &= -4\pi\ddot{P}, \\
\ddot{P} + 2\gamma_2\dot{P} + (\omega_0^2 + \gamma_2^2)P &= -\frac{2d^2\omega_0}{\hbar}N\mathscr{E}, \\
\dot{N} + \gamma_1(N - N_0^1) &= \frac{2\mathscr{E}}{\hbar\omega_0}(\dot{P} + \gamma_2 P).
\end{aligned}\right\} \tag{1}$$

These equations are a logical generalization of the equations for a molecular oscillator [56-59]. They have been derived in the more recent literature, for example, in [32, 60]. In the equations, $\mathscr{E}$ is the electric field strength in the resonator, P is the average value of the polarization of the active particles, N is the difference between the number of particles at the upper and lower levels, N_0^1 is the initial difference, $\omega_0 = (w_2 - w_1)/\hbar$ is the transition frequency, ω_r is the frequency of the resonator, Q is the quality of the resonator, $\gamma_2 = 1/\tau_2$, $\gamma_1 = 1/\tau_1$ (where τ_1 and τ_2 are the relaxation times of the diagonal and off-diagonal elements of the density matrix), d is the modulus of the dipole moment matrix element. We seek the solution of the system (1) in the form (see, e.g., [61])

$$\mathscr{E} = E_1(t)\cos[\omega_0 t + \varphi(t)], \quad P = P_1(t)\cos[\omega_0 t + \psi(t)], \tag{2}$$

where E_1, P_1, φ, and ψ are slowly varying time functions.

It is postulated that N(t) also varies slowly with the time. Application of the Van der Pol method to the system (1) yields the following system of equations for the slowly changing variables [34, 37]:

$$\left.\begin{aligned}
\frac{dx}{dt} &= -\gamma_3 x + \gamma_3 y \sin_1\Phi, \\
\frac{dy}{dt} &= -\gamma_2 y + \gamma_2 xz \sin_1\Phi, \\
\frac{dz}{dt} &= -\gamma_1(z - z_0) - \gamma_1(z_0 - 1)xy\sin_1\Phi, \\
\frac{d\Phi}{dt} &= -B + \left(\gamma_2\frac{xz}{y} + \gamma_3\frac{y}{x}\right)\cos_1\Phi.
\end{aligned}\right\} \tag{3}$$

The following dimensionless variables have been introduced here: $x = E_1/E_{10}$, $y = P_1/P_{10}$, $z = N/N_0$, as well as the notation $\sin_1\Phi = \sin\Phi/\sin\Phi_0$, $\cos_1\Phi = \cos\Phi/\sin\Phi_0$, $B = (\omega_r^2 - \omega_l^2)/2\omega_0$, $(\omega_l^2 = \omega_0^2 + \gamma_2^2)$, $\gamma_3 = \omega_0/2Q$, where the subscript zero refers to the stationary values of the variables $(E_{10}, P_{10}, N_0, \Phi_0)$, $\Phi = \psi - \varphi$ is the phase difference between the polarization and the field. The stationary values for the case B = 0 have the form

$$E_{10} = \sqrt{4\pi Q\hbar N_0^1\gamma_1}, \quad P_{10} = \sqrt{\frac{N_0^1\hbar\gamma_1}{4\pi Q}}, \quad N_0 = \frac{\hbar\gamma_2}{4\pi Qd^2}, \quad \Phi_0 = \frac{\pi}{2}. \tag{3a}$$

2. Rate Equations

Equations (1) and the corresponding equations (3) are rigorous equations for the description of a two-level laser (within the framework of the τ_1, τ_2 model), taking into account the diagonal and off-diagonal elements of the density matrix. Often in the literature, however, lasers are described by means of the rate equations (equations including only the diagonal elements of the density matrix).

Conditions for Applicability of the Rate Equations (within the Framework of the τ_1, τ_2 Model). Let us see under what conditions the system (3) goes over to the rate equations [35], which have the form [29]

$$\left.\begin{array}{l} \dfrac{d}{dt} E_1^2 = -2\gamma_3 E_1^2 + C\omega_0 N E_1^2, \\[2mm] \dfrac{d}{dt} N = -\gamma_1 (N - N_0^1) - D N E_1^2. \end{array}\right\} \tag{4}$$

The notation here is the same as in the preceding section; C and D were introduced in [29] as arbitrary constants. For a comparison of the systems (3) and (4) we write the latter in dimensionless variables and reduce it to a second-order differential equation:

$$\ddot{x} - \frac{\dot{x}^2}{x} + \frac{\omega}{\gamma_3}\dot{x}x^2 + \frac{\gamma_1}{\omega}\dot{x} + x(x^2 - 1) = 0, \tag{5}$$

where the differentiation is carried out with respect to the time $\tau = \omega t$, and ω is determined from the relation

$$\omega^2 = D E_{10}^2 \gamma_3. \tag{6}$$

The constant C was included in the stationary value of E_{10}. On the other hand, as will become evident below [38], it may be strictly said that for B = 0 in Eqs. (3), $\cos \Phi \equiv \cos \Phi_0 \equiv \pi/2$. With this fact in mind, the system (3) may be written as one third-order equation, namely

$$\frac{\omega}{\gamma_2 + \gamma_3}\left\{\dddot{x} - \frac{\ddot{x}\dot{x}}{x}\right\} + \ddot{x} - \frac{\dot{x}^2}{x} + \frac{\omega}{\gamma_3}\dot{x}x^2 + \frac{\gamma_1}{\omega}\dot{x} + x(x^2 - 1) = 0, \tag{7}$$

where the differentiation is carried out with respect to the time $\tau = \omega t$, and ω is determined from the relation

$$\omega^2 = \frac{d^2}{\hbar^2(\gamma_2 + \gamma_3)} E_{10}^2 \gamma_3. \tag{8}$$

From a comparison of (6) and (8) we obtain $D = d^2/\hbar^2(\gamma_2 + \gamma_3)$.

It is apparent that Eq. (7) goes over to Eq. (6) on fulfillment of three conditions:

$$\frac{\omega}{\gamma_2 + \gamma_3} \ll 1, \tag{9a}$$

$$\frac{\omega}{\gamma_2 + \gamma_3} \ll \frac{\gamma_1}{\omega}, \tag{9b}$$

$$\frac{\omega}{\gamma_2 + \gamma_3} \ll \frac{\omega}{\gamma_3}. \tag{9c}$$

The first condition (9a) is obvious, while the conditions (9b) and (9c) are necessary in order to justify the presence of the terms $(\gamma_1/\omega)\dot{x}$ and $(\omega/\gamma_3)\dot{x}x^2$ in the equation. The condition (9c) is fulfilled if $\gamma_2 \gg \gamma_3$ or

$$\frac{1}{\tau_2} \gg \frac{\omega_0}{2Q}. \tag{10}$$

This condition was deduced in [62] and implies that the width of the spectral line (equal to $1/\tau_2$) is considerably larger than the linewidth of the resonator. However, in addition to (9c), there are still the conditions (9a) and (9b). Both of these impose a restriction on the active particle flux or the magnitude of the field in the resonator. Since $\tau_1 > \tau_2$, the condition (9b) is stronger [provided (10) is fulfilled]. It may be written in the form

$$\frac{d^2 E_{10}^2}{\hbar^2} \ll \gamma_1 \gamma_2 \frac{\gamma_2}{\gamma_3} . \tag{11a}$$

If we introduce the ratio η of the number of active particles to the limiting number of active particles, the condition (11a) assumes the form

$$\eta - 1 \ll \frac{\gamma_2}{\gamma_3} . \tag{11b}$$

Consequently, within the framework of the τ_1, τ_2 model the rate equations are valid when the conditions (10), (11a), and (11b) are fulfilled, i.e., when the linewidth is much larger than the resonator width and the field in the resonator is not too large (the excess of η over the threshold value is less than γ_2/γ_3).

Character of Nonstationary States in the Rate Equations. The problem of nonstationary conditions and their character within the framework of the rate equations has been investigated in many papers [39, 40, 43-45, 53, 54]. In [39, 40] the most rigorous proof is given for the fact that the stationary solution is always stable in such equations and that under no circumstances does the pulsing condition set in. Only the stabilization process is possible in the rate equations, i.e., any perturbed solution ultimately recovers its stationary value. Following [36], we seek to explain the nature of the process whereby the system (4) is restored to the stationary state. We write (4) in the somewhat alternative form

$$\left.\begin{aligned}
\frac{dN}{dt} &= b\mathscr{P} - \frac{N}{\tau_1} - \frac{\alpha}{\tau_1} nN, \\
\frac{dn}{dt} &= \frac{\alpha}{\tau_1} nN - 2\gamma_3 n.
\end{aligned}\right\} \tag{4a}$$

Here the term $\gamma_1 N_0^1$ in the second equation of the system (4) is written in the form $b\mathscr{P}$, where b is some coefficient, $\mathscr{P}$ is the pumping power, i.e., the pumping is incorporated in explicit form, rather than through the initial number of active particles in the resonator. In place of the field density in the resonator we have introduced the number of photons n, and the coefficient α/τ_1 is equivalent† to D. Equating the derivatives in Eqs. (4) to zero, we obtain the stationary values:

$$N_0 = \frac{2\gamma_3 \tau_1}{\alpha}, \quad n_0 = \frac{b\mathscr{P}}{2\gamma_3} - \frac{1}{\alpha}$$

and the threshold value of the power:

$$\mathscr{P}_{\text{thresh}} = \frac{2\gamma_3}{\alpha b} .$$

Let us suppose that the system (4a) has undergone a small perturbation, i.e., that $N = N_0 + \Delta N$, $n = n_0 + \Delta n$, where ΔN and Δn are deviations from the stationary state, $\Delta N \ll N_0$ and $\Delta n \gg n_0$. Then, linearizing the system (4a) in the vicinity of the stationary values, we obtain the following equation for Δn:

$$\frac{d^2}{dt^2}(\Delta n) + \frac{\mathscr{P}}{\mathscr{P}_{\text{thresh}} \tau_1} \frac{d}{dt}(\Delta n) + \frac{2\gamma_3}{\tau_1} \left(\frac{\mathscr{P}}{\mathscr{P}_{\text{thresh}}} - 1 \right) \Delta n = 0. \tag{12}$$

This represents the equation for a damped harmonic oscillator. Its solution is well known (see,

---

† It is at once evident that Eqs. (4a) coincide with Eqs. (4) if we let $E_{10}^2 = \theta n$ and $C\omega_0 = D\theta$.

e.g., [63]):

$$\Delta n = \{A \cos \omega t + B \sin \omega t\} \exp\left(-\frac{\mathscr{P}t}{2\mathscr{P}_{\text{thresh}}\tau_1}\right), \tag{13}$$

where

$$\omega^2 = \frac{2\gamma_3}{\tau_1}\left(\frac{\mathscr{P}}{\mathscr{P}_{\text{thresh}}} - 1\right), \tag{13a}$$

and A and B are determined from the initial conditions.

If ω (13a) is larger than the exponent of the exponential function, the number of photons Δn returns to the equilibrium state, pulsating and dying out; if the exponent is greater than ω, the exponential defines the solution (13) and Δn decays almost exponentially without pulsating. It is interesting that the damping of (13) is accelerated when the Q of the resonator is increased and the pumping power is increased (see Introduction). In fact, the exponent may be written in the form

$$\frac{\mathscr{P}}{2\mathscr{P}_{\text{thresh}}\tau_1} = \frac{1}{2\tau_1}(1 + \alpha n_0) = \frac{1}{2\tau_1}\frac{\alpha b \mathscr{P}}{2\gamma_3}.$$

The systems of Eqs. (4) and (4a) are the simplest. Often in the literature one sees more complex modifications of the rate equations [12, 44, 45], but the results of their investigation are the same: 1) The rate equations do not have nonstationary solutions; 2) the arrival at a stationary state is quicker, the higher the Q of the resonator and the larger the pumping power; conversely, the transition to a pulsing (or very slowly damped) state must occur with a reduction in the Q and pumping power.

Updating of the Rate Equations. A number of authors have proposed modernizations of the rate equations (4) and (4a). The introduction of additional terms therein results in the appearance of terms corresponding to negative friction in Eq. (12), the onset of instability, and a pulsing state. It is suggested in [3, 10] that the damping constant γ_3 be assumed to depend on the resonator field or, equivalently, on the number of active particles. It is proposed in [39, 43, 47] that an additional term be incorporated into the equation for the population difference. However, the physical essence of the proposed terms remains unclear at this juncture.

3. States in a System of Equations Including Off-Diagonal Elements of the Density Matrix

It is possible in the rigorous system (3) for a pulsating state to make its appearance [33, 34].

Case B = 0 (No Frequency Separation between the Resonator and Line). An analysis of the stability of the stationary state of the system (3) leads to the following instability condition [33, 48] for the stationary state:

$$z_0 > \frac{\gamma_3(\gamma_3 + \gamma_1 + 3\gamma_2)}{\gamma_2(\gamma_3 - \gamma_1 - \gamma_2)}. \tag{14}$$

A condition similar to (14) for the case $\tau_1 = \tau_2$, i.e., $\gamma_1 = \gamma_2$, has been derived in [49, 50]. The condition (14) means that the stationary harmonic state becomes unstable for a sufficiently large initial number of active particles in the resonator. It is important to note that the inequality (14) can only be satisfied with the stipulation that

$$\gamma_3 > \gamma_1 + \gamma_2. \tag{15}$$

Otherwise the inequality reverses, and it cannot be satisfied for any values of z_0 ($z_0 > 0$). An investigation of the condition (14)† at the extremum with respect to γ_3 shows that if

$$\gamma_3 = (\gamma_1 + \gamma_2)\left(1 + \sqrt{2\,\frac{\gamma_1 + 2\gamma_2}{\gamma_1 + \gamma_2}}\right), \tag{15a}$$

then fulfillment of the condition (14) requires a minimum number of active particles. If $\gamma_1 \ll \gamma_2$, (15a) goes over to the relation $\gamma_3 = 3\gamma_2$, i.e., the optimum conditions for the onset of a pulsating state are realized when the linewidth of the resonator is three times the width of the spectral line.

Let us make some numerical estimates of the condition (14). We consider a molecular oscillator [56, 57] for this purpose. Typical values of the parameters for this device are: $\gamma_1 = \gamma_2 = 10^3$ cps, $\nu = 24 \cdot 10^9$ cps, $d = 1.8 \cdot 10^{-18}$, typical active particle fluxes are 10^{12} particles per sec. For these values the condition (14) is satisfied with a resonator Q of $5 \cdot 10^4$. Direct application of the system (3) and the condition (14) to lasers (ruby) interposes certain difficulties. If we consider the τ_1, τ_2 model to be valid for ruby, then $\gamma_2 = 3 \cdot 10^{11}$ cps and $\gamma_3 = 10^8$ to 10^9 cps for that material, i.e., the condition (15) is never fulfilled and the pulsing state does not occur. This difficulty could possibly be circumvented within the framework of a multimode model [37] or by the formulation of a new theory accounting for the shape and width of the ruby lines [34].

We seek the solution to (3) in the region of instability [34], i.e., for fulfillment of the condition (14). Substituting the first and second equations of the system (3) into the last equation of that system, we obtain

$$\frac{d}{dt}\ln(xy\cos\Phi) = -(\gamma_2 + \gamma_3). \tag{16}$$

Equation (16) is readily integrated. The solution has the form

$$\cos\Phi = \frac{C}{xy}\,e^{-(\gamma_2 + \gamma_3)\,t}, \tag{17}$$

where C is a constant depending on the initial conditions. It is clear that whatever these conditions happen to be, in the steady state $\cos\Phi \equiv 0$, i.e., $\Phi_1 \equiv \Phi_0 \equiv \pi/2$. Consequently, with exact alignment of the resonator to the frequency of the line in any nonstationary state, the phase difference between the polarization and field is constant and equal to $\pi/2$ [38]. If we make use of this result, the system (3) goes over to the third-order system of equations:

$$\left.\begin{aligned}
\frac{dx}{dt} &= -\gamma_3 x + \gamma_3 y,\\
\frac{dy}{dt} &= -\gamma_2 y + \gamma_2 xz,\\
\frac{dz}{dt} &= -\gamma_1(z - z_0) - \gamma_1(z_0 - 1)xy.
\end{aligned}\right\} \tag{18}$$

Introducing the new time $\tau = \omega_1 t$ (we will define ω_1 below) and the small parameters $\mu = \omega_1/\gamma_3 \ll 1$, $\mu_1 = \gamma_1/\omega_1 \ll 1$, and $\mu_2 = \gamma_2/\gamma_3 \ll 1$, we reduce the system (18) to a third-order nonlinear equation [34, 50] of the form

$$\ddot{x} - \frac{\dot{x}^2}{x} + \frac{\gamma_1\gamma_2(z_0 - 1)}{\omega_1^2(1 + \mu_2)}\,x(x^2 - 1) = -\mu\left[(1 + \mu_2)^{-1}\left(\dddot{x} - \frac{\ddot{x}\dot{x}}{x}\right) + \mu_1(1 + \mu_2)\,\ddot{x} - \frac{\mu_1}{\mu}\,\dot{x} + \frac{\gamma_1\gamma_2(z_0 - 1)}{\omega_1^2(1 + \mu_2)}\,x^2\dot{x}\right]. \tag{19}$$

---

† For an investigation at the extremum the condition (14) must be used in the form (26).

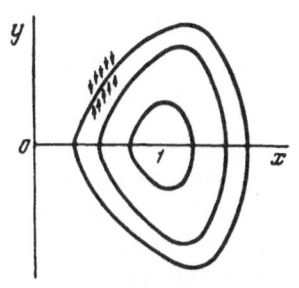

Fig. 1. Family of cycles in the phase plane.

The introduction of the parameter μ_2 is closely allied with the condition (14) (assuming, of course, that $\gamma_1 \ll \gamma_2$). It is at once apparent from Eq. (19) that it is convenient† to choose

$$\omega_1^2 = \frac{\gamma_1\gamma_2(z_0-1)}{1+\mu_2} = \frac{d^2E_{10}^2}{\hbar^2(1+\mu_2)},$$ (20)

where we have made use of the expression (3a) for the stationary values. Then correct to first-order terms in μ, Eq. (19) assumes the form

$$\ddot{x} - \frac{\dot{x}^2}{x} + x(x^2-1) = -\mu\left(\dddot{x} - \frac{\ddot{x}\dot{x}}{x} + \frac{\mu_1}{\mu}\dot{x} + \dot{x}x^2\right).$$ (21)

This is a nonlinear equation with a small nonlinearity on the right-hand side. It is solvable by standard methods [64, 65]. For $\mu = 0$ the equation is integrable. We obtain

$$\dot{x} = \pm x\sqrt{C + 2\ln x - x^2}.$$ (22)

The solution (22) defines a family of cycles in the Van der Pol plane, with C as a parameter. This family is shown in Fig. 1. The motion along the cycle is stable, as indicated by the arrows in the figure. The point x = 1 corresponds to the stationary solution $E_1 = E_{10}$. If we expand ln x about the point x = 1, we quickly obtain an explicit expression for the field amplitude as a function of the time:

$$E_1 = E_{10}\frac{2 - C_1}{2 - \sqrt{2C_1}\sin\left[\frac{dE_{10}}{\hbar}\sqrt{2-C_1}(t+h_+)\right]}.$$ (23)

Here h_+ is an arbitrary constant related to the suitability of the time chosen as a time reference, and $C_1 = C - 1$. For the period we obtain a value

$$T = \frac{2\pi}{\frac{dE_{10}}{\hbar}\sqrt{2-C_1}}.$$ (24)

In Eqs. (23) and (24) we have reverted to ordinary time $t = \tau/\omega_1$, and ω_1 is taken correct to first order in μ_2. Determining the constant C_1 is a fairly straightforward though tedious process. It is accomplished by standard methods [64, 65]. The final result (correct to first order in μ) has the form

$$C_1 = \frac{2\left[1 - \frac{\gamma_1\gamma_3}{\left(\frac{dE_{10}}{\hbar}\right)^2}\right]}{3 - \frac{1}{4}\frac{\gamma_1\gamma_3}{\left(\frac{dE_{10}}{\hbar}\right)^2}}.$$ (25)

A few words are in order regarding the solution (23)-(25). As yet the condition (14) has not been satisfied, there are no cycles, no solution (3a). On fulfillment of the condition (14), pulsations appear against the background of the constant average amplitude E_{10}. Their amplitude is still small, and they are nearly sinusoidal in shape, but then harmonics begin to appear.

_____

† See also Eq. (8).

The pulsing frequency is proportional to the average field amplitude and increases with the amplitude. The mechanism of the process [34] consists in the rapid transfer of particles from the upper to the lower level with the liberation of energy (ascending portion of the spike) and the rapid ejection of particles by the field from the lower to the upper level with the absorption of energy (descending portion of the spike).

Fulfillment of the condition (14) implies that the number of particles in the resonator is fairly large, the luminescence energy exceeds the losses in the resonator, and the energy in the resonator increases. This in turn increases the probability of luminescence, resulting in still further growth of the field. However, the number of particles is diminished in this process, and eventually the influx of energy becomes equal to the losses. Now there are a large field and some of the particles at the lower level in the resonator. The field energy can either emerge from the resonator or eject some of the particles back again to the upper level. On fulfillment of the condition (14) the second process is the more likely. Consequently, the large field impels the particles upward again, and the field is diminished. This is the descending part of the spike. Then the whole thing starts all over again. The pulsing frequency is of the order $dE_{10}/\hbar$. This is precisely the changeover frequency of the particles from level to level [66]. Pulsing occurs with a high saturation when every active particle goes from the upper to the lower level and back again several times during the period of interaction with the field. Due to the inverted population, transition from the upper level always prevails on the average. This is also the case, of course, in the pulsing state (the field amplitude never drops to zero). The condition (14) may also be written in the form

$$N_0^1 > \frac{\hbar}{4\pi d^2 Q} \frac{\gamma_3(\gamma_3 + \gamma_1 + 2\gamma_2)}{\gamma_3 - \gamma_1 - \gamma_2}. \tag{26}$$

If the condition (26) is not fulfilled, then increasing the Q of the resonator (recalling that $\gamma_3 = \omega_0/2Q$) results in fulfillment of the condition (26), i.e., the onset of undamped pulsations. The opposite conclusion is inferred from the rate equations. Increasing the pumping power results in the onset of pulsing, within the framework of the given analysis, but, as shown in [50], it is to be expected that pulsing will vanish with increasing pumping power when a third level is included.

Case of Nonzero Frequency Separation between the Line and the Resonator in the System (3). Let us consider the system (3) without assuming precise tuning of the resonator to the line frequency (B ≠ 0). The instability of such a system is investigated in a manner similar to that of the perfectly aligned system. The final result is far too cumbersome to be computed in general form. This condition is of the same type as the instability condition of the harmonic state in a molecular oscillator [67]. We write this condition for two extreme cases. On fulfillment of the inequality $\gamma_3 \gg \gamma_2 \gg \gamma_1$ we obtain the following instability condition for the case of a small frequency separation B < γ_3:

$$z_0 > \frac{\gamma_3}{\gamma_2}\left[1 + 4\left(\frac{B}{\gamma_3 + \gamma_2}\right)^2\right]. \tag{27}$$

For the case of large separation $B/\gamma_3 \gg 1$, the instability condition has the form

$$z_0 > \frac{\gamma_3}{\gamma_2} 2 \frac{\gamma_3}{\gamma_1}. \tag{28}$$

The number of active particles (proportional to z_0) required in order for instability to set in is much larger in the case of large separation than for small separation. An abrupt rise in the critical number of particles occurs in the region $B/\gamma_3 \sim \frac{1}{2}$, i.e., for a separation of half the resonator width the condition for the onset of a nonstationary state deteriorates appreciably. It can also be shown (the computations involved are elementary but cumbersome) that with frequency separation present, as in the case of perfect alignment, fulfillment of the condition $\gamma_3 > \gamma_2 + \gamma_1$ is essential for the onset of the nonstationary state; see (14).

We have seen that the phase Φ [system (3)] is constant for zero separation and that only the amplitude is modulated with the onset of pulsing [solution (23)]. In the presence of frequency separation the phase is no longer constant, but pulsates, resulting in additional amplitude modulation with a frequency approximately equal to the frequency separation. We will demonstrate this (see [38]). Let us consider the system (3) for the case of sufficiently large B ($B \gg \gamma_3$) and seek a more rapid modulation than (23)-(25), i.e., with frequencies ω satisfying the conditions

$$\omega \gg \omega_n, \; \omega \gg \gamma_3, \tag{29}$$

where $\omega_n = 2\pi/T$ [see (24) regarding T].

We solve the system (3) for this case. The last equation in (3) may be brought to the form

$$\frac{d}{dt}[\tan\Phi] = -B\tan\Phi + (\gamma_3 + \gamma_2) + \frac{d}{dt}\ln\left[x\left(x + \frac{\dot{x}}{\gamma_3}\right)\right]. \tag{30}$$

By the method of variation of constants we obtain the solution of Eq. (30):

$$\tan\Phi = \frac{x\,(x + \dot{x}/\gamma_3)}{Be^{-(\gamma_2+\gamma_3)t}\int x\,(x + \dot{x}/\gamma_3)\,e^{(\gamma_2+\gamma_3)t}\,dt}. \tag{31}$$

Introducing the small parameters $\mu = \gamma_3/\omega$ and $\mu_1 = \gamma_2/\gamma_3$ [see the conditions (29)], changing to the dimensionless time $\tau = Bt$, and performing some straightforward but rather tedious manipulations, we obtain from (31) the following equation for the phase difference Φ as a function of τ and $\xi = x^2$:

$$\tan\Phi = \frac{\dot{\xi}}{\xi + C_1}\{1 + \mu F_1\}, \tag{32}$$

where

$$F_1 = \frac{2\dot{\xi}}{\xi} + \frac{G_1}{\xi + C_1} - \frac{\int(\xi - C_1)\,dt}{\xi + C_1}, $$

and the following equation for ξ as a function of τ:

$$\frac{\ddot{\xi}}{\xi} - 2\frac{\ddot{\xi}\dot{\xi}}{\xi^2} + \frac{\dot{\xi}^3}{\xi^3} + C_1^2\frac{\dot{\xi}}{\xi^3} = \mu F_2, \tag{33}$$

where

$$F_2 = -\frac{\dddot{\xi}}{\xi} + \frac{\dot{\xi}^2}{\xi} + \frac{C_1}{\xi} - \frac{C_1^2}{\xi^2} - \frac{\gamma_1\gamma_2\,(z_0 - 1)}{\mu\omega^2\sin^2\Phi_0}\dot{\xi} + 4\frac{\mu_1}{\mu}C_1^2\frac{\dot{\xi}}{\xi^3} - \frac{2C_1\dot{\xi}\int(\xi - C_1)\,dt}{\xi^3}. $$

We assumed in the derivation of Eqs. (32) and (33) that ξ is a periodic time function and, consequently, may be expanded in a Fourier series:

$$\xi = C_1 + \sum_{n \neq 0} A_n e^{in\omega t}. \tag{34}$$

The nonlinear equation (33) is solvable. For $\mu = 0$, integrating twice, we obtain the equation of a cycle in the phase plane:

$$\dot{\xi} = \pm\frac{1}{\sqrt{|C_2|}}\sqrt{-\xi^2 + \frac{C_3}{|C_2|}\xi - \frac{C_1^2}{|C_2|}}. \tag{35}$$

One more integration yields the solution

$$x = \sqrt{\frac{C_3}{2|C_2|} \pm \frac{\sqrt{-\Delta}}{2|C_2|} \sin\left[B\sqrt{|C_2|}\,(t-t_0)\right]},$$ (36)

the variables in this case being x and $t = \tau/\omega$, where

$$\Delta = 4C_1^2|C_2| - C_3^2 < 0; \quad \frac{C_3}{2|C_2|} = C_1 \text{ [see eq. (34)]}.$$

The constants $|C_2|$ and C_3 are determined by standard methods [64, 65]. They are easily found if the first integrals of the nonlinear equation (33) are known for $\mu = 0$. For the constant $|C_2|$ we obtain the algebraic equation

$$|C_2|^2(|C_2| - 1) = \frac{\pi}{8}\left(\frac{\gamma_3^2}{B\gamma_2}\right)^2.$$ (37)

This cubic equation is easily solved if the right-hand side is small. Then one of the roots of the equation is nearly equal to unity. In this case the final result for the frequency of the process becomes

$$\omega = B\left[1 + \frac{1}{4}\left(\frac{\pi}{8}\frac{\gamma_3^2}{B\gamma_2}\right)^2\right].$$ (38)

With an increase in B and γ_2 the frequency tends to B. The quantity α is the ratio of the variable to the constant component in the solution (36), given the same assumptions as those under which the frequency was determined, and is equal to the following:

$$\alpha = \frac{\sqrt{-\Delta}}{C_3} = \frac{1}{2}\frac{\pi}{8}\frac{\gamma_3^2}{B\gamma_2}.$$ (39)

For the phase difference Φ between the polarization and field we substitute the solution (36) into (32) to obtain the following correct to zeroth order in μ:

$$\tan\Phi = \frac{\alpha\omega\cos\omega\,(t - t_0)}{2 + \alpha\sin\omega\,(t - t_0)},$$ (40)

where ω is determined by Eq. (38) and α by Eq. (39). It is apparent that Φ executes oscillations about the equilibrium position $\Phi = 0$, which corresponds to the stationary state (in fact, if we include first-order small terms, the stationary value turns out to be $\tan\Phi_0 = \gamma_3/B \ll 1$ and the phase oscillates about this stationary value). The phase oscillations lead to amplitude modulation of the field (36). This represents beats between the polarization of the medium and the resonator field. They occur in the region of instability and cause the particles in the laser to be emitted either "more weakly" or "more strongly."

The maximum emission will occur when the phase difference is largest, the minimum emission when the phase difference is a minimum (the phase difference tending to zero corresponds to infinite separation). Consequently, within the framework of the two-level system (the same equations are equally valid for a four-level system) the stationary harmonic state is unstable in some interval of values determined by the condition (14), and pulsations occur; in this event two amplitude modulation mechanisms are possible, each having a different frequency and different percentage modulation. One mechanism is given by the solution (23)-(25), the other by (36), (38)-(40).

4. Stabilization Processes in a Two-Level Laser

Equations (1) and their corresponding equations of the system (3) enable one not only to describe stationary processes in a laser, but also the stabilization or settling processes therein

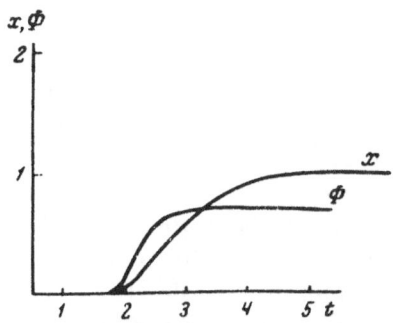

Fig. 2. Buildup of the amplitude x and phase Φ of a molecular oscillator for $\tau_f \ll \tau$ (arbitrary scale).

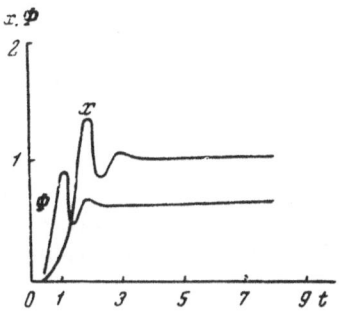

Fig. 3. Buildup of the amplitude x and phase Φ of a molecular oscillator for $\tau_f \approx \tau$ (arbitrary scale).

[33]. For simplicity we analyze the characteristics of the stabilization processes in a molecular oscillator [56, 57], i.e., for the case $\gamma_1 = \gamma_2$. For the molecular oscillator [33] the stabilization (relaxation) time $1/\gamma_3$ of the resonator oscillations is much less than the transit time τ_1 of the molecules through the resonator. This means that during the initial time interval immediately following activation of the molecular oscillator the quantities x and Φ will vary rapidly (in the phase $\Phi = \psi - \varphi$, the component φ varies while ψ remains constant), while the quantities y and z will be essentially invariant. Consequently, the amplitude of the oscillations in the resonator will grow rapidly (this occurs during a period on the order of $1/\gamma_3$), then follows the variation of the polarization and number of active particles in quasi-continuous fashion. After the oscillator is turned on (we impose on the sorting system a voltage sufficient for self-excitation of the oscillator), oscillation does not begin immediately. The oscillator has a certain "dead time" T_m, which is determined by the time required for a sufficient number of molecules to accumulate for realization of the self-excitation condition and the time for growth of the signal amplitude to the noise level of the receiving device. It can be shown [33] that

$$T_m = \tau_f \left(1 + \ln \frac{\eta}{\eta - 1} \right) + \frac{\tau_1}{\eta - 1} \ln \frac{x_n^2}{\chi x_i^2}, \qquad (41)$$

where x_n is the amplitude of the receiver noise, χ is the power coupling coefficient of the resonator and receiver, x_i is the initial amplitude of the oscillations in the resonator and is determined by the spontaneous noise or thermal fluctuations in the resonator, η is the self-excitation coefficient, and τ_f is the time for formation of the active particle beam in the system and depends on the input voltage to the sorting system.

It is apparent from (41) that T_m depends strongly on η (η is a function of the number of active particles), i.e., on the intensity of the active particle beam. An investigation of the transient processes in the linear approximation shows that the stabilization time of the oscillations in a molecular oscillator depends on the self-excitation coefficient η and the beam forming time τ_f. Figures 2 and 3 show typical time dependences of the amplitude and phase of a molecular oscillator for $\tau_f \ll \tau$ and $\tau_f \approx \tau$; the curves were generated on a digital computer. For $\tau_f \approx \tau$ the curves have a clearly pronounced oscillator character, for $\tau_f \ll \tau$, the stabilization processes are monotonic with respect to time. An increase in η produces a steeper rise in the oscillation amplitude at first, but then the rate of approach to the stationary state after the first spike is almost independent of η. Analytical investigation leads to the following damped solution in the linear approximation [33]:

$$\Delta n = (\Delta n_0) e^{-t/2\tau_1} \cos \left[(\eta - 1.25)^{1/2} \frac{t}{\tau_1} + \bar{\varphi} \right]. \qquad (42)$$

Consequently, the stabilization time in the system is equal in order of magnitude to the transit time τ_1 through the resonator, and the frequency $\omega = \sqrt{(\beta - 1.25)/\tau_1}$, i.e., it depends on the number of active particles.

5. Other Hypotheses Pertinent to the Pulsing Theory

Attempts have been made in several studies to relate the pulsing of laser emission to noise. The first attempt in this direction was made in [31], where the influence of spontaneous emission on the nonstationary states of a quantum oscillator were investigated. It was shown that if the noise is large, it can exert a powerful influence on such states (the state of regular damped pulsations assumes a clearly pronounced irregularity, becoming chaotic). In another paper [51], the influence of the pumping, treated as a stochastic process, on the amplitude and phase of electromagnetic oscillations is examined within the scope of a single mode. It is shown that self-excitation of the laser occurs for sufficiently high pumping levels, and the amplitude of the light oscillations tends to some stationary value. Allowance for the statistical properties of the pumping imbues the stabilization processes themselves with a significantly fluctuating character and causes the amplitude to execute small fluctuations with the following amplitude about the equilibrium position in the steady state:

$$\rho = \sum_{\substack{t \\ \mu\nu}}^{t} R_{\mu\nu} K\,(t,\,t_{\mu\nu}),$$

where

$$K\,(t\sigma) = \frac{1}{2i\,\sqrt{2}}\,[e^{i\omega_{\pi}(t-\sigma)} - e^{i\omega_{2}(t-\sigma)}]; \quad \omega_{\pi,\,2} = \frac{i\alpha}{2} \pm \sqrt{-\frac{\alpha^2}{4} + 2\sigma};$$

$$R_{\mu\nu} = \mathrm{Re}\,[e^{-i\varphi}ih^{*}\delta\alpha_{\mu}^{+}\,(t_{\mu\nu})];$$

$\alpha = \gamma + k$ (k characterizes the resonator losses, γ the probability of spontaneous emission); α_{μ}^{+} is the creation operator, corresponding to particle transition under the influence of pumping in the excited state:

$$\delta\alpha_{\mu}^{+}\,(t_{\mu\nu}) = \alpha_{\mu}^{+}\,(0)\,e^{i\varphi_{\mu\nu}} - \alpha_{\mu}^{+}\,(t_{\mu\nu} - 0);$$

φ is the phase of the light wave; $h = -\frac{e}{m}\,\sqrt{\frac{2\pi}{\hbar\omega V}}\int \varphi_{1} n p \varphi_{2}\,d\tau$, where V is the volume of the resonator, p is the momentum operator, and φ_{1} and φ_{2} are the wave functions of the electron states of the atom.

The influence of noise (spontaneous emission) on the emission pattern of lasers seems to be important, in general, but it scarcely offers the sole ultimate explanation of the pulsing state (certainly, at any rate, not of the regular pulsing condition). The tendency to relate the pulsing to the presence of several oscillatory modes definitely bears watching (postulates 3 and 4 in the Introduction). The first efforts are just now being made in this direction, and it is too early to offer any final conclusions. The pulsing problem is treated in [12] within the framework of a multimode model. The rate equations are used to describe the problem. For a population difference N(zt)

$$\frac{dN\,(zt)}{dt} = \gamma_{1}\,[N\,(zt) - N_{0}^{1}] - \sum_{i} Dg_{i}N\,(zt)\,P_{i}\,(zt),$$

where N_{0}^{1} is the initial value, D and g_{i} are constants, i is the mode number, γ_{1} is the probability of spontaneous emission, and

$$P_{i}\,(zt) = N_{i}\,(t)\left(1 - \cos\frac{2m_{i}\pi}{L}\,z\right).$$

For the number of photons in the resonator

$$\frac{dn_i}{dt} = -\gamma_i n_i + D g_i n_i \int_0^L N(zt)\left(1 - \cos\frac{2m_i\pi}{L} z\right) dz.$$

It is clear that the presence of a spatial distribution is only accounted for in the direction of the z axis; L is the length of the resonator, m_i is the number of half-waves that fit the resonator length.

The above system of equations was solved on a computer for the case of two modes. Damped pulsation states were obtained. The authors arrived at the following conclusions. 1) The sum of the amplitudes of the two modes executes regular damped oscillations, while the amplitude of each mode executes irregular oscillations; 2) the amplitudes in the two different modes pulsate in antiphase, i.e., when the amplitude of one phase is a maximum the amplitude of the other is a minimum, and vice versa. The stability of the stationary state has been investigated in [52] for the case of two modes. Although the equations were more rigorous than the rate equations, it still turned out that a nonstationary state would not arise in the system.

LITERATURE CITED

1. T. H. Maiman, R. H. Hoskins, I. J. D'Haenens, C. K. Asawa, and V. Evtuhov, Phys. Rev., 123:1151 (1961).
2. R. J. Collins, D. F. Nelson, A. L. Schawlow, W. Bond, C. G. B. Carrett, and W. Kaiser, Phys. Rev. Letters, 5:303 (1960).
3. M. D. Galanin, A. M. Leontovich, Z. A. Sviridenkov, V. N. Smorchkov, and Z. A. Chizhikova, Opt. i Spektroskopiya, 14:165 (1963).
4. T. P. Hughes, Nature, 195(4839):325 (1962).
5. A. J. de Maria and R. Gagosz, Appl. Opt., 2(8):807 (1963).
6. P. Walsh and G. Kemeny, J. Appl. Phys., Part 1, 34(4):956 (1963).
7. Z. A. Chizhikova, M. D. Galanin, V. V. Korobkin, A. M. Leontovitch, and V. N. Smortchkov, Quantum Electronics, p. 1483. Proc. Third Internat. Congress, Columbia University Press, New York (1964).
8. S. Koosekanani, M. Giftan, and A. Krutchkoff, Appl. Opt., 1(3):372 (1962).
9. M. Shimazu, I. Ogura, A. Hashimoto, and H. Sasaki, Proc. Symposium on Optical Masers, p. 405. Polytech. Press, New York (1963).
10. K. Shimoda, Proc. Symposium on Optical Masers, p. 95. Polytech. Press, New York (1963).
11. B. J. McMurtry and A. E. Siegman, Appl. Opt., 1(1):51 (1962).
12. C. L. Tang, H. Statz, and G. de Mars, J. Appl. Phys., 34:2289 (1963).
13. K. Gürs, Quantum Electronics, p. 1113. Proc. Third Internat. Congress, Columbia University Press, New York (1964).
14. Kiroshi Takuma, Japan. J. Appl. Phys., 2:197 (1963).
15. E. Snitzer, Quantum Electronics, p. 999. Proc. Third Internat. Congress, Columbia University Press, New York (1964).
16. A. K. Sokolov and T. N. Zubarev, Dokl. Akad. Nauk SSSR, 159:539 (1964).
17. V. K. Konyukhov, L. A. Kulevskii, A. M. Prokhorov, and A. K. Sokolov, Dokl. Akad. Nauk SSSR, 158:824 (1964).
18. C. M. Stickley, Proc. IEEE, 51:848 (1963).
19. C. M. Stickley and R. L. Townsend, Jr., Proc. Symposium on Optical Masers, p. 541. Polytech. Press, New York (1963).
20. W. R. Mallory, Proc. IEEE, 51:850 (1963).
21. G. L. Clark, R. F. Wuerker, and C. M. York, J. Opt. Soc. Am., 52:878 (1963).

22.   M. Silver, R. S. Witte, and C. M. York, Appl. Opt., 3(4) : 539 (1964).

23.   D. Röss, Proc. IEEE, 51 : 468 (1963).

24.   R. E. Johnson, W. H. McMahan, F. J. Oharek, and A. P. Sheppard, Proc. IRE, 49 : 1942 (1961). (1961).

25.   V. G. Veselago and Yu. V. Kosichkin, Radiotekhn. i Elektron., 8(6) : 967 (1963).

26.   P. P. Sorokin, J. H. Stevenson, J. R. Lankard, and G. D. Pettit, Phys. Rev., 127 : 503 (1962).

27.   D. F. Nelson and W. S. Boyle, Appl. Opt., 1(2) : 181 (1962).

28.   W. R. Bennet, Phys. Rev., 126 : 580 (1962).

29.   H. Statz and G. de Mars, Quantum Electronics, p. 530. Columbia University Press, New York (1960).

30.   J. R. Singer and S. Wang, Phys. Rev. Letters, 6 : 351 (1961).

31.   A. V. Gaponov and V. I. Bespalov, Izv. Vysshikh. Uchebn. Zavedenii Radiofiz., 8 : 70 (1965).

32.   Yon-Han Pao, J. Opt. Soc. Am., 52 : 871 (1962).

33.   A. Z. Grasyuk and A. N. Oraevskii, Radiotekhn. i Elektron., 9(3) : 524 (1964).

34.   V. V. Korobkin and A. V. Uspenskii, Zh. Eksperim. i Teor. Fiz., 45 : 1003 (1963).

35.   V. N. Genkin and Ya. N. Khanin, Izv. Vysshikh. Uchebn. Zavedenii Radiofiz., 5 : 423 (1962).

36.   H. A. Bostick and J. R. O'Connor, Proc. IRE, 50 : 219 (1962).

37.   A. Z. Grasyuk and A. N. Oraevskii, Estratto da Rendiconti della Scuola Internationale di Fisica "E. Fermi," XXXI Corso.

38.   A. V. Uspenskii, Radiotekhn. i Elektron., 9(4) : 747 (1964).

39.   D. M. Sinnet, J. Appl. Phys., 33 : 1578 (1962).

40.   K. N. Rozhdestvenskii, Radiotekhn. i Elektron., 8(12) : 2043 (1963).

41.   J. E. Kaplan and R. Zier, J. Appl. Phys., 33 : 2372 (1962).

42.   E. J. Post, Appl. Opt., 1(2) : 165 (1962).

43.   G. Makhov, J. Appl. Phys., 33 : 202 (1962).

44.   A. M. Ratner, Zh. Eksperim. i Teor. Fiz., 45 : 1908 (1963).

45.   D. A. Ostrovskii and E. I. Yakubovich, Zh. Eksperim. i Teor. Fiz., 46 : 963 (1964).

46.   M. Birnbaum, T. Stocker, and S. J. Wells, Proc. IEEE, 51 : 854 (1963).

47.   G. Makhov and O. Risgin, Quantum Electronics, p. 1121. Proc. Third Internat. Congress. Columbia University Press, New York (1964).

48.   J. I. Kaplan, J. Appl. Phys., 34 : 3411 (1963).

49.   A. S. Gurtovnik, Izv. Vysshikh. Uchebn. Zavedenii Radiofiz., 1(5-6) : 83 (1958).

50.   A. V. Uspenskii, Radiotekhn. i Elektron., 8(7) : 1165 (1963).

51.   H. Haken, Preprint.

52.   H. Haken and H. Sauermann, Z. Phys., 173 : 261 (1963).

53.   R. Dunsmuir, J. Electron. Control, 10 : 453 (1961).

54.   R. W. Hellworth, Phys. Rev. Letters, 6 : 9 (1961).

55.   J. E. Lundman, Appl. Opt., 2(8) : 862 (1963).

56.   N. G. Basov and A. M. Prokhorov, Usp. Fiz. Nauk, 7 : 485 (1955).

57.   N. G. Basov, Doctoral Dissertation, FIAN AN SSSR (1956).

58.   A. N. Oraevskii, Radiotekhn. i Elektron., 4 : 718 (1959).

59.   V. M. Fain, Zh. Eksperim. i Teor. Fiz., 33 : 945 (1957).

60.   L. W. Davis, Proc. IEEE, 51 : 76 (1963).

61.   Kh. Yu. Khaldre and R. V. Khokhlov, Izv. Vysshikh. Uchebn. Zavedenii Radiofiz., 1(5-6) : 60 (1958).

62.   G. L. Tang, J. Appl. Phys., 34 : 2935 (1963).

63.   A. A. Andronov, A. A. Vitt, and S. É. Khaikin, Theory of Oscillations. Fizmatgiz, Moscow (1959).

64.   I. G. Malkin, Certain Problems in the Theory of Nonlinear Oscillations. Gostekhizdat, Moscow (1956).

65.   V. A. Volosov, Usp. Mat. Nauk, 17(6) : 20 (1962).

66.   L. D. Landau and E. M. Lifshits, Quantum Mechanics. Fizmatgiz, Moscow (1963).

67.   V. N. Lugovoi, Izv. Vysshikh. Uchebn. Zavedenii Radiofiz., 7 : 792 (1964).

INVESTIGATION OF THE GAS LASER

N. G. Basov, É. M. Belenov, E. P. Markin,
V. V. Nikitin, and A. N. Oraevskii

INTRODUCTION

The purpose of the present article is to explore the possibility of enhancing the power output of various gas lasers by a proper choice of optimum laser operating conditions (pressure of the gas mixture, partial pressures of the components, pumping power, mirror transmissivity, diameter and length of the discharge tube).

In view of the capability of generating a whole series of modes at high gains, it is pertinent to investigate the beam divergence and spectrum of the generated radiation as a function of the laser power output. We have studied these problems from the experimental and theoretical standpoint simultaneously.

The theory provides an explanation for the experimentally observed optimum discharge parameters, for which peak laser power is observed. The experimental values obtained for the optimum parameters of the lasers were used as a basis for the design of a high-power gas laser.

Using a neon−helium mixture, it was possible to obtain a power of 100 mW at a wavelength $\lambda = 1.15\,\mu$ with an optimum tube radius of 8 mm and length of 3 m.

The angular and modulation characteristics were studied as a function of the output power. It turned out that the output power influences the angular spread of the light flux emitted by the laser, this being attributable to the appearance of additional oscillatory modes. The separation of one given mode should eliminate the increase in the angle of divergence associated with elevated power.

The investigation of the modulation characteristics shows that a localized frequency modulation of approximately 10^5 cps can be attained by means of internal modulation.

The investigation of the spectral characteristics of the gas laser showed clearly that the most sensitive method for their measurement is the method of the rotating laser. An apparatus for measuring the emission spectrum of a gas laser was designed and assembled on the basis of that method.

The connection of several lasers in series by means of a prism, as well as the use of a discharge tube of rectangular cross section, one dimension of which is optimum, offers a considerable increase in the power output of the gas laser.

CHAPTER I

CONSTRUCTION OF THE LASER

Extensive investigations on the operation of the laser and elucidation of the influence exerted by various parameters on the output power of the laser stimulated the problem of designing a gas laser whose construction would be the most workable and reliable in operation. The distinctive feature of the laser we built compared with those described earlier [1] is the capability of interchanging discharge tubes of various diameters (from 2 to 60 mm) and lengths without altering the position of the mirrors or their alignment, thus facilitating the observance of identical conditions throughout an experiment.

The construction of the laser is shown in Fig. 1.

It is seen in the figure that the mirror 1 is placed in a steel collar 2, where it is held in place by a cover nut 3. A protective quartz plate 4 is mounted in the cover 5, which is attached to the collar 2 by four screws. A vacuum seal is ensured by the rubber insert 6. The collar 2 is attached by two flexible steel screws to the base 7, the screws having proved to be more effective than springs. The mirror is rotated by means of a micrometer screw 8 through the lever mechanism 9, which makes it possible to set the mirrors parallel to one another, correct to within 0.5".

The rotating mechanism is constructed such that the deviation of the mirror in alignment occurs in mutually perpendicular planes. The flanges 7 are interlinked by four steel rods (D = 25 mm, l = 1 m, the length may be varied), which imparts the necessary rigidity to the construction. The gas discharge tube 10 does not have a soldered joint with the chamber in which the mirror is situated, but is attached by means of a Wilson seal 14 and is isolated from the base 7 by stainless steel sylphons 11. Pumping and ignition of the gas are accomplished through the inlet ports 12 and 13.

A Fabry−Perot interferometer incorporating high-reflectance mirrors (R ~ 99%, T ~ 0.1%), produced by the vacuum deposition of alternating quarter-wave dielectric films with different refractive indices, serves as the resonator in the laser. It is known that a thin homogeneous layer deposited on a glass plate contained between two semitransparent parallel interface surfaces also acts as a Fabry−Perot etalon, for which the maximum reflection coefficient is com-

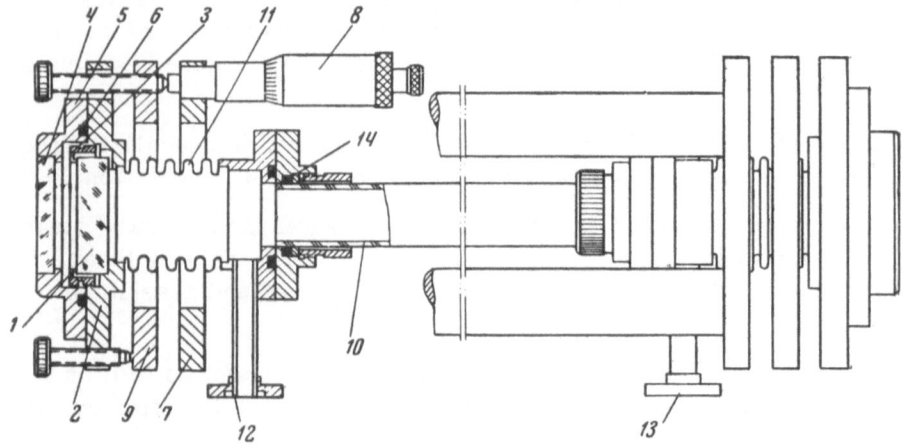

Fig. 1. Construction of the gas laser.

puted from the equation [2]

$$R_{\max} = \left(\frac{n_2 - n_1^2}{n_2 + n_1^2} \right)^2,$$

where n_1 is the refractive index of the film and n_2 is the refractive index of the backing. The equation is valid with the stipulation that the minimum thickness of the layer $t = \lambda/4$.

It is evident from the equation that R increases with n_1. In practice, only a few substances have refractive indices ~3, and those available have a high absorption coefficient. It is known from experimental investigations [2, 3] that the best materials in the sense of low absorption are ZnS (n = 2.3), Sb_2S_3 (n ~ 3), TiO_2 (n = 2.4), and others. One layer of ZnS reflects ~30% of the incident light [2]. In order to attain high values of R, it is necessary either to choose materials with large n or to alternate layers of different substances with $t = \lambda/4$, so as to ensure a maximum disparity between the refractive indices ($n_2 - n_1$). In our work we used mirrors either with dielectric coatings or of metal. In our first generation experiments we used dielectric mirrors. In order to conduct detailed investigations of the relationship between the laser power output and the variation of the mirror parameters (transmissivity T, pass band Δf, reflection coefficient R, etc.) for different wavelengths, we developed and constructed spray apparatus with precision optical monitoring of the thickness of the dielectric layers, thus permitting the use of very high-quality mirrors.

After achieving the required vacuum of ~10^{-3} mm Hg, the system was supplied with a mixture of spectrally pure gases from a buffer volume, in which the gases were mixed in the necessary proportions. The gas mixture was fed to the gas discharge tube from the buffer volume through a needle injector. The total pressure in the laser was recorded by a MacLeod oil manometer calibrated in mm Hg. This arrangement permitted the gas mixture to be replenished in the laser on the one hand, the pressure in the tube to be varied rather simply on the other. The vacuum just prior to ignition of the gas was ~10^{-6} mm Hg, so as to have a completely sealed-off system.

A glow discharge was excited in the tube by a high-frequency generator (f = 28 Mc), the frequency of which was stabilized by a quartz cavity. The construction of the pumping generator provided for smooth variation of the power from 0 to 300 W. The laser was matched to the discharge tube by variation of the inductive coupling.

The design provided for the inclusion of external amplitude modulation either by pulsed or by sinusoidal oscillations with a frequency from 50 cps to 1 Mc.

The laser output was recorded as follows. The emergent beam from the laser was directed to an interference filter (or the slit of an IKS-12 monochromator) and was focused by means of a quartz lens onto a sensing area (photoresistance, photodiode, photomultiplier).

For amplification of the signal received by the sensing unit a narrowband resonance amplifier was used, tuned to the beam chopper frequency (f = 186 cps). The amplifier had the following parameters: gain K ~ 10^4, pass band Δf ~ 4 cps, sensitivity of several microvolts.

An electric motor with disc was used to modulate the output beam of the laser. By varying the voltage on the motor, it was possible to smoothly regulate the chopper frequency between 160 and 200 cps. The amplified signal was sent to the input of an oscilloscope. The onset of generation was signaled on the scope screen by the appearance of sinusoidal oscillations with the chopper frequency and on the screen of the electron-optical converter by the appearance of bright luminescence.

Before undertaking the task of producing a beam with a mixture of neon and helium, preliminary investigations were conducted on the spectra of the pure neon and helium components, as well as their mixture in various proportions. This was done by focusing the illumination of the gas discharge (from the end of the tube with one mirror removed) on the slit of an IKS-12 monochromator and recording the spectrum on an ÉPP-09 spectrometer using a synchronous detector. It was found that the radiation intensity at a wavelength $\lambda = 1.15\ \mu$ increases in going from pure neon to the mixture. Also investigated was the influence on the radiation intensity at $\lambda = 1.15\ \mu$ of the pressure of the mixture, the pumping power, etc., whereupon the optimum conditions were established for emission at the given wavelength.

For the purpose of studying different gas mixtures and learning new generatable wavelengths, we built a laser with confocal external mirrors [4]. This type of laser is highly suitable because of the low diffraction losses incurred [5] (the low losses in a confocal resonator are attributable to the strong field concentration near the interferometer axis). The discharge tube was hermetically sealed by means of quartz plates 5 mm thick, mounted at the Brewster angle with respect to the laser axis. These plates do not introduce reflection losses for radiation polarized in the plane of incidence. The only losses occur as the result of scattering and absorption in the plates. The tolerance on the Brewster angle is about 2-3°, thus permitting the laser to operate over a wide spectral range. Inasmuch as the mirrors are external, there is no danger of their malfunctioning from the action of the discharge or due to chemically active gases, and they can be rapidly exchanged to fit the proper wavelength.

The discharge tube is held by means of three riders on an optical bench. The mirrors are installed in a head with spring-mounted alignment screws. Tungsten leads are provided at the ends of the discharge tube for the dc excitation of discharge.

Plates made of LK-5 or K-8 glass were mounted at the Brewster angle or methane gas was injected into the cavity during operation of the laser at $\lambda = 6328$ Å for partial suppression of the competing $\lambda = 3.39\ \mu$.

The mirrors were aligned in the rotating laser experiment by means of the red line $\lambda = 6328$ Å. This experiment will be described below.

Experiments were performed on the laser with external mirrors to study the internal modulation, as well as the influence of additions of Ar, Kr, N_2, CO_2 gases on the output power.

CHAPTER II

INVESTIGATION OF THE DEPENDENCE
OF THE LASER OUTPUT POWER ON VARIOUS PARAMETERS

1.   Methods of Measuring the Laser Output Power

The output power of the laser was measured by transmitting its beam through an interference filter and focusing it through a lens onto the sensing area of a photoresistance (the power was measured in relative units per mV on the instrument photoresistance). First of all the dependence of the voltage U at the receiver output on the light beam intensity I for the corresponding wavelength was investigated. The measurements indicate a linear dependence for the characteristic $U = f(I)$. This method of measuring the laser power permits the characteristics to be recorded with ease and rapidity and then its optimum operating conditions to be ascertained.

The absolute power of the laser was measured in the optimum state by two independent techniques: a) by means of a special photodiode calibrated in terms of the radiant energy of the sun and in terms of a thermopile; b) by means of a calorimeter. The photodiode was calibrated

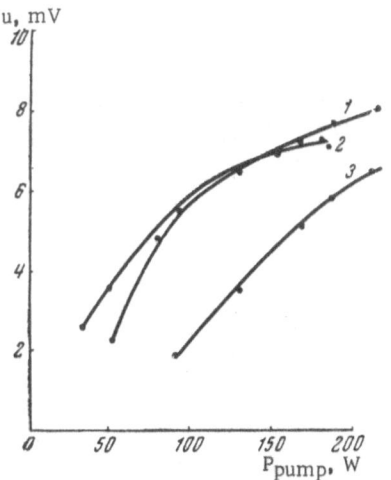

Fig. 2. Dependence of the laser output power in relative units (mW) on the pumping power at various pressures of the gas mixture and for a tube 6 mm in diameter ($\lambda = 1.15\ \mu$). 1) p = 1.5; 2) p = 2.0; 3) p = 0.71 mm Hg.

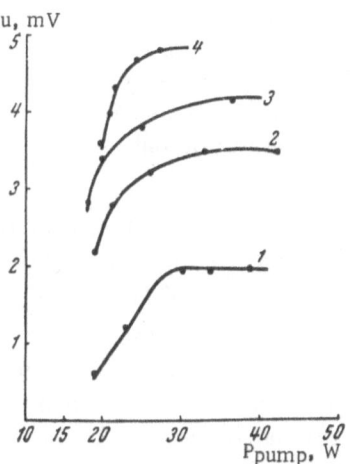

Fig. 3. Dependence of the laser output power on the pumping power at various pressures of the gas mixture for a tube 10 mm in diameter ($\lambda = 1.15\ \mu$). 1) p = 0.4; 2) p = 0.6; 3) p = 0.8; 4) p = 1.0 mm Hg.

according to the thermopile as follows. The light from the heat source was focused on the thermopile so that its entire working area was illuminated. The required wavelength was separated by an interference filter.

Once the constant of the thermopile was known, the power of the light flux was calculated. Then, without changing the experimental conditions, the light flux was measured with the photodiode. Direct measurement of the laser output power with the thermopile could introduce sizable errors due to the nonuniform density of the output beam over its cross section. The density of the beam varies over the cross section as a function of the quality of the mirrors and the operating conditions of the laser (presence of various modes). Among the shortcomings of the photodiode is its selectivity in a certain region of the spectrum (0.6-2 μ). Consequently, in order to measure the power in any region of the optical spectrum, we made a calorimeter of the "rat's nest" type [6]. The calorimeter is made up primarily of bolometers, which comprise two Dewars filled with copper enameled wire with d = 0.05 mm, R = 1 kΩ. Its ends are connected in a bridge circuit. The other two arms of the bridge are a 1-kΩ resistance and an MSR-28 resistance box. The emission from the laser is sent to the working bolometer once the circuit is balanced. The balance of the bridge is upset due to variation of the wire resistance as the result of heating.

If the resistance of the wire prior to irradiation was R_0, after irradiation it increases to $R = R_0(1 + \alpha\Delta\theta)$, where α is the temperature coefficient of the resistance, $\Delta\theta$ is the change in temperature of the wire. Then the change in resistance is equal to

$$\Delta R = R_0\alpha\,\Delta\theta.$$

By measuring ΔR, it is possible to find $\Delta\theta$, and hence the energy absorbed by the wire, which is directly proportional to the change in resistance and is computed from the relation

$$E = k\,\Delta R.$$

The readings of our instrument were compared with those of the calorimeter described in [7], which was calibrated in terms of the radiated power of a ruby laser. The agreement was quite good.

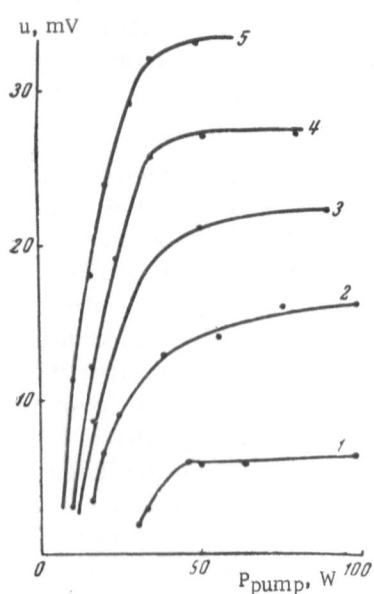

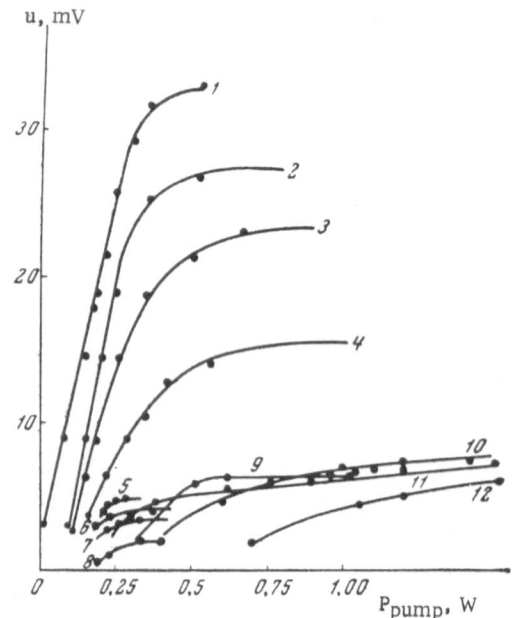

Fig. 4. Dependence of the laser output power on the pumping power at various pressures of the gas mixture for a tube 8 mm in diameter ($\lambda = 1.15\,\mu$). 1) p = 0.3; 2) p = 0.5; 3) p = 0.7; 4) p = 1.0; 5) p = 1.2 mm Hg.

Fig. 5. Dependence of the laser output power on the pumping power for tubes with diameters of 6 (curves 9-12), 8 (1-4), and 10 mm (5-8) ($\lambda = 1.15\,\mu$) at various pressures of the gas mixture (mm Hg). 1) p = 1.2; 2) p = 1.0; 3) p = 0.7; 4) p = 0.5; 5) p = 1.0; 6) p = 0.8; 7) p = 0.6; 8) p = 0.4; 9) p = 0.3; 10) p = 1.5; 11) p = 2.0; 12) p = 0.7 mm Hg.

2. Characteristics of the Laser Operating on a Mixture of Ne and He ($\lambda = 1.15\,\mu$)

In [1, 8-10] an emitted power of 4-15 mW ($\lambda = 1.153\,\mu$) was realized with a Ne−He gas laser. The present investigation was undertaken with the object of studying the influence of various parameters (pumping power, diameter and shape of the discharge tubes, discharge length, pressure of the gas mixture, transmissivity of the mirrors) on the output power of a Ne−He gas laser [11].

To investigate the dependence of the beam power on the diameter of the tubes, we fabricated the laser whose construction is described above (Fig. 1). We used plane interference-type reflecting mirrors with quarter-wave dielectric coatings having different transmission coefficients (T ~ 0.2, 2, 8%).

The discharge was excited in quartz tubes by a high-frequency generator (f = 28 Mc), the power output of which could be varied from 0 to 500 W. Spectrally pure Ne and He were used for the experiments. In all the tests the partial pressure of the Ne and He was 1:9. The power measurements were carried out for the transition $2S_2 - 2P_4$ in Ne (wavelength λ = 11,530 Å).

The graph in Fig. 2 shows the dependence of the laser output power on the pumping power for different gas pressures and a tube 6 mm in diameter. The laser power is plotted (in relative units) on the vertical axis, the pumping power on the horizontal. As the graph indicates, the largest output power is obtained at a pressure p ~ 1.5 mm Hg and a pumping power P_{pump} ~ 200 W. The discharge length was 750 mm. The maximum power (absolute) in these tests for the mirrors with T ~ 0.2% was ~0.4 mW. Although the output power continued to increase with

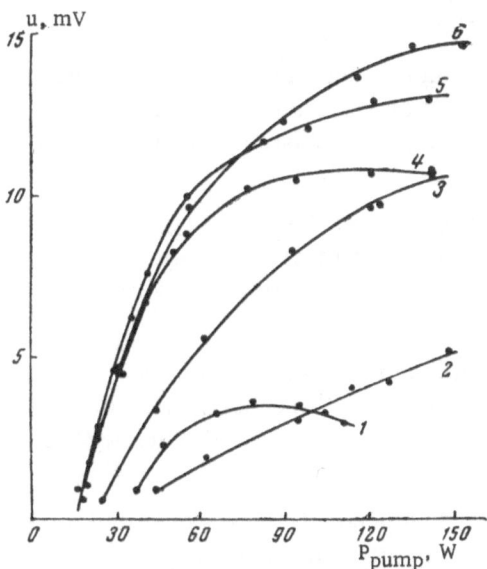

Fig. 6. Dependence of the laser output power on the pumping power at various pressures of the gas mixture for a tube of rectangular cross section ($\lambda = 1.15$ μ). 1) p = 3.0; 2) p = 0.5; 3) p = 0.8; 4) p = 2.0; 5) p = 1.5; 6) p = 1.1 mm Hg.

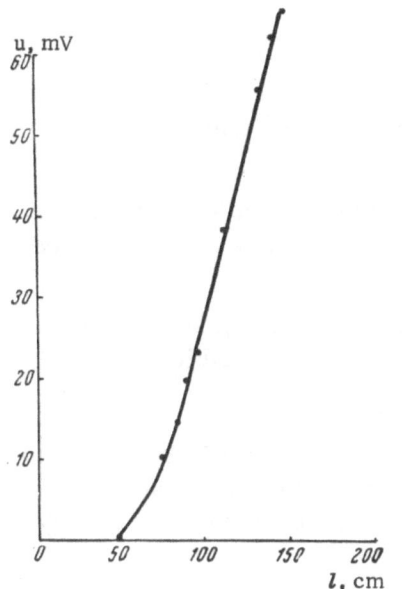

Fig. 7. Dependence of the laser output power on the discharge length ($\lambda = 1.15$ μ).

increasing pumping power, we were unable to raise the latter higher than 200 W, because this resulted in breakdown of the dielectric coatings of the mirrors due to the discharge.

An analogous graph is shown in Fig. 3 for the tube 10 mm in diameter. As apparent from the figure, the laser power increases almost linearly with the pumping power up to some definite value ($P_{pump} \sim$ 30-50 W), after which it remains almost invariant. The maximum value of the output power in this case was obtained at a pressure p $\sim$ 1 mm Hg and a pumping power $P_{pump} = 30$ W.

The dependence of the laser output power on P_{pump} for the 8-mm tube is shown in Fig. 4. At a pressure p $\sim$ 1.2 mm Hg the laser power (absolute) reaches a maximum ($\sim$1.6 mW) at a pumping power of $\sim$50 W.

Curves showing the dependence of the laser output power on P_{pump} for the 6-, 8-, and 10-mm tubes with a discharge length of 750 mm are presented on the same scale in Fig. 5 for comparison.

It is apparent from these curves that the laser utilizing a tube 8 mm in diameter gives a four- and sevenfold gain in power over the 6- and 10-mm tubes, respectively, for a discharge length of 750 mm. Clearly, a tube diameter of 8 mm is nearly optimal for a Ne−He mixture, although such an acute variation in the laser power due to small changes in the tube diameter seems strange indeed. It was observed in studying the dependence of the generated power on the shape of the discharge tube that the power was increased by using a rectangular tube having one inside dimension of $\sim$8 mm and the other arbitrary. In the test laser we used a discharge tube with a cross section 8×20 mm^2 and lengths of 1 and 3 m. The laser exhibited stable operation and yielded a power of $\sim$1 mW. Using the same mirrors and a cylindrical tube (20 mm in diameter), the power was exceedingly small and the laser operation was highly unstable. The gain in power using a discharge tube with rectangular cross section (8×20 mm^2) over a round tube 8 mm in diameter proved to be 2- to 2.5-fold. Curves for the dependence of the generated power on the pumping power are shown for the rectangular tube in Fig. 6, whence it follows that the pumping power does not change appreciably with a cylindrical tube. A considerable increase is to be expected in the output power if a rectangular tube is used.

An investigation of the dependence of the laser output power on the discharge length was performed using a laser with a 300-cm separation between the mirrors.

We investigated the dependence of the generated power on the pumping power for a tube 10 mm in diameter and a discharge length of 1 and 2 m, as well as the analogous curves for a tube 8 mm in diameter. The maximum power was obtained at a gas pressure of 1 mm Hg. Generation cutoff occurred at P_{pump} = 250 W and p = 4 mm Hg.

The graph in Fig. 7 shows the dependence of the output power on the discharge length under near-optimal conditions ($P_{pump} \sim$ 200 W, p $\sim$ 1.2 mm Hg, mirror transmission coefficient T $\sim$ 0.20%). As apparent from the graph, the laser power increases linearly with the discharge length, beginning with $l \sim$ 0.6-0.7 m.

For the first time the dependence of the generated power on the transmission coefficient of the mirrors (T $\sim$ 0.2, 2, 8%) was investigated. The maximum power was observed for mirrors with T $\sim$ 2%; for mirrors with T $\sim$ 0.2% it was smaller by $1/10$, while for mirrors with T $\sim$ 8% it was smaller by $1/30$.

It is interesting that the generated power is essentially independent of the quality of finish of the mirror backings. For mirrors whose backings were machined with a precision of 3λ, practically the same power was observed (glass backings were also used) as had been observed for mirrors machined with a precision of 0.03λ, although the beam structure observed on the screen of the electron-optical converter depended strongly on the quality of finish of the backing surfaces.

For a laser with a discharge length of 2m, generation was realized at the following wavelengths: 1.145, 1.153, 1.160, 1.165, and 1.170 μ.

For a discharge length of 1 m, the same laser produced a beam at two wavelengths: 1.153 and 1.160 μ.

The experimental investigations permitted certain conclusions to be drawn regarding possible techniques for enhancing the output power of an Ne−He laser:

1. It was established that a tube diameter close to the optimum (cavity with plane mirrors) is D = 8 mm.

2. The laser output power increases linearly as a function of the discharge length, beginning with $l \sim$ 0.6 or 0.7 m.

3. The optimum discharge parameters have been determined, including: the pumping power, total pressure of the mixture in the discharge tube, partial pressures of the Ne and He.

4. An investigation of the dependence of the laser output power on the transmissivity of the mirrors for a definite resonator length has made it possible to enhance the output power significantly.

These data were used as a basis for designing and building a laser with the following parameters: discharge tube diameter 8 mm, resonator length 300 cm, discharge length 2.80 cm, plane mirrors with alternating quarter-wave dielectric coatings of ZnS and cryolite with a transmissivity T $\sim$ 2%.

During the experiment the total pressure of the mixture was held at $\sim$1.2 mm Hg. The output power of this laser at λ = 1.153 μ, corresponding to transition from the $2S_2$ to the $2P_4$ level in Ne, was equal to 100 mW. Using another laser design with spherical mirrors (confocal resonator), a power of $\sim$150 mW was obtained (discharge length 320 cm, resonator length 350 cm).

In the given experiment a tube with an inside diameter of 12 mm was used.

The disparity between the optimum discharge tube diameters for plane-parallel and confocal mirror configurations is explained by the fact that the gain of the active medium per unit length plays the dominant role in the plane mirror case, whereas, in the spherical mirror case this is true of the distance between the mirrors and the radiation wavelength [12].

Investigations of the dependence of the laser output power (plane resonator) on the diameter of the discharge tube showed that the optimum diameter is equal to 8 mm.

We started with lasers consisting of two and six discharge tubes mounted in parallel, each one 8 mm in diameter (several tubes bunched together) and 1 and 2 m in length, operating with the same plane mirrors. All the tubes were filled simultaneously with Ne and He.

The anticipated results with respect to increased power are given by the rectangular tubes we investigated. The use of several circular or rectangular tubes in a bunch, or the mounting of trihedral prisms in place of the mirrors made it possible to increase the laser output power considerably. The length of the discharge increases in this case.

The principal factor limiting the length of the resonator is diffraction. The diffraction losses increase with the length L as $L^{3/2}$, so that the quality of the resonator is given by the following relation, taking diffraction into account:

$$Q = \frac{2\pi L}{\lambda} \frac{1}{(1-r) + \lambda_{mn}^2 \frac{4.62}{\pi \sqrt{\pi}} \left(\frac{L\lambda}{D^2}\right)^{3/2}},$$

where λ_{mn} is the nth root of the mth-order Bessel function

$$I_m(\lambda_{mn}) = 0,$$

r is the reflection coefficient of the mirrors $(1-r \ll 1)$, λ is the wavelength, and D is the diameter of the discharge tube.

The condition for excitation of the laser in terms of the negative absorption coefficient η may be written in the form

$$\eta = \frac{\lambda}{2\pi} kQ > 1.$$

For good operation of the laser it is desirable that η be at least 3-5, since the laser efficiency increases with Q.

The condition $(\lambda/2\pi)kQ \sim 3\text{-}5$ may be used to find the optimum length of the equipment, which turns out in excitation of the fundamental mode (m = 0, n = 1, λ_{01} = 2.4) for $\lambda = 1\,\mu$, k = 0.15 m^{-1}, D = 1 cm, to be equal to 20-30 m.

For a large negative absorption coefficient η it increases as k^2.

Consequently, there is a possibility of increasing the power of existing gas lasers by another order of magnitude, using bunched discharge tubes.

3. Characteristics of the Laser Operating on a Mixture of Ne and He ($\lambda = 3.39\,\mu$)

In the present section we describe some experimental investigations of the dependence of the output power of a Ne−He laser on various parameters for the $3S_2-3P_4$ line, $\lambda = 3.39\,\mu$ [13], which has a high gain (~160%/m [14]).

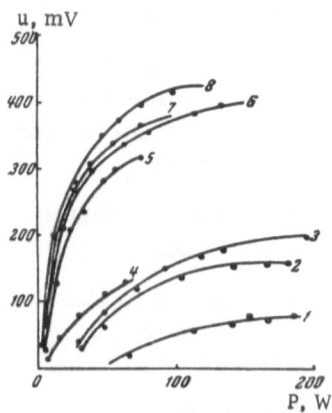

Fig. 8. Dependence of the laser output power ($\lambda = 3.39\ \mu$) on the pumping power at various pressures of the gas mixture. 1) p = 6.0; 2) p = 5.0; 3) p = 4.0; 4) p = 0.3; 5) p = 0.5; 6) p = 1.5; 7) p = 2.7; 8) p = 1.2 mm Hg.

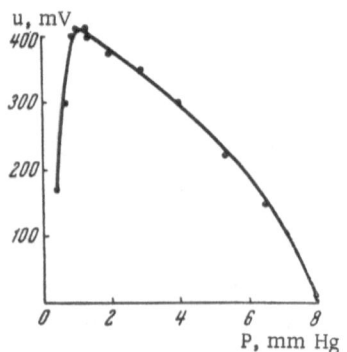

Fig. 9. Dependence of the laser output power on the pressure of the gas mixture ($\lambda = 3.39\ \mu$).

The curves in Fig. 8 show the dependence of the laser output power in relative units (millivolts on the instrument photoresistance) on the pumping power at various pressures for a discharge tube 12 mm in diameter. The discharge length is 0.9 m. The receiver in this case comprised liquid nitrogen-cooled photoresistances of PbSe and InSb. As apparent from the graph, the maximum laser power is observed at a pumping power of ~100 W and p ~ 1.2 mm Hg. The lower generation threshold is equal to 3 W. The diameter of the generated spot was equal to the diameter of the discharge tube. The laser exhibited stable operation with the pressure of the gases in the tube varied from 0.3 to 6 mm Hg; at a pressure of ~8 mm Hg, generation cutoff took place.

In recording the characteristics, we used plane reflecting mirrors with a silver coating having transmission coefficients of 2 and 0.1%. These characteristics were recorded for tubes with diameters of 8, 10, 15, and 20 mm, and it was established from an analysis of the curves that the near-optimum diameter with respect to power output is D = 12 mm.

Figure 9 shows the dependence of the laser output power on the pressure of the mixture in the discharge tube. It is evident from the figure that the power depends slightly on the pressure variation between 0.9 and 1.4 mm Hg.

It was also found that the output power increases linearly with the length of discharge. It is interesting to note that the laser exhibited stable operation for a discharge length of 2 or 3 cm and a pumping power of several watts. This indicates the conceivability of a small-scale laser operating at $\lambda = 3.39\ \mu$.

We investigated the dependence of the laser output power on the transmission coefficient of the mirrors (T ~ 2, 30, 50%). The maximum power was observed for mirrors with T ~ 30%.

For a laser 300 cm long operating under optimum conditions (p ~ 1 mm Hg, P_{pump} ~ 100 W, transmission coefficient of the interference mirror T ~ 30%) a maximum power of ~300 mW was obtained.

We note that the addition of a third component to the Ne—He mixture, namely spectrally pure Xe in small fractions, on the order of a few percent, led to the observation of simultaneous generation at the following wavelengths (plane mirrors with silver coating): $\lambda = 1.52, 2.02, 2.60, 3.10, 3.36, 3.39,$ and $3.50\ \mu$, along with others but much weaker in intensity. In working with a Xe—He mixture, we observed generation (intense lines) at $\lambda = 2.02, 2.60, 3.10, 3.36,$ and $3.50\ \mu$, which concurs with the results given in [12].

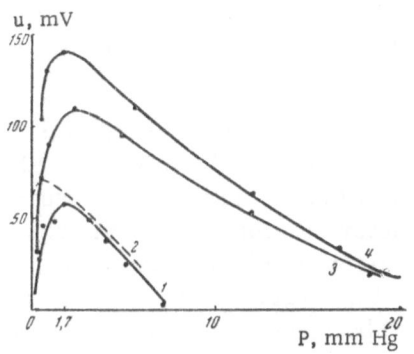

Fig. 10. Dependence of the output power of the Xe−He laser on the gas pressure. 1) $\lambda = 3.50\mu$, d = 20 mm; 2) $\lambda = 3.36\mu$, d = 8 mm; 3) $\lambda = 3.50\mu$, d = 12 mm; 4) $\lambda = 3.50\mu$, d = 8 mm.

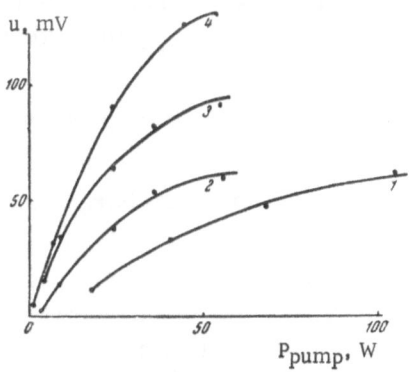

Fig. 11. Dependence of the laser output power on the pumping power for a tube 12 mm in diameter. 1) $\lambda = 3.50\mu$, p = 11 mm Hg; 2) $\lambda = 3.36\mu$, p = 0.13 mm Hg; 3) $\lambda = 3.50\mu$, p = 0.13 mm Hg; 4) $\lambda = 3.50\mu$, p = 2.09 mm Hg.

It was observed that when one of the mirrors was gradually deflected out of parallel, oscillations appeared in the output power (about 100 peaks for a 1° deflection of the mirror). The experiment was set up as follows. The emergent beam from the laser was sent to a PbSe photoresistance cooled with liquid nitrogen. The dimensions of the sensing area were 2×2 mm². In order to suppress the intensity of spontaneous emission, the receiver was placed at a distance 0.5 m from the laser and was mounted on a special stand with a micrometer drive for displacement of the photoresistance in the vertical and horizontal directions. The laser output power was measured in relative units (in millimeters amplitude of the signal observed on the oscilloscope screen).

Prior to the measurements, the laser mirrors were aligned in parallel according to the maximum generated power, for which purpose each mirror was aligned in turn. Then, by smoothly deflecting the mirror from parallel, the power oscillations on either side of the maximum were fixed. Hence, with a decrease in generated power (by variation of P_{pump} or the gas pressure in the tube) the number of oscillations decreases sharply, and with a further reduction in laser power it is possible to realize an operational state in which a single maximum is observed within the limits of a small $\alpha = \pm 7\text{-}10''$.

Since $\lambda = 3.39\mu$ cannot be recorded on photographic film or by an image converter, we used the following method to study the intensity of the laser beam over its cross section. We used a photoresistance with a small sensing area (1×2 mm²), which was displaced by micrometer screws throughout the cross section of the laser beam, correct to 0.01 mm. It was established that the beam diameter depends on the operating conditions of the laser and varies from 10 to 12 mm. The effective cross section of the tube was 12 mm.

4. Characteristics of the Laser Operating on a Mixture of Xe and He ($\lambda = 3.50\mu$)

In [12, 15] the generated wavelengths are given for a xenon−helium laser; it was found that the most intense line was the Xe $3d_4 - 2p_9$ line with $\lambda = 3.50\mu$; the gain of this line is equal to 50 dB [15].

In the present study we investigated the dependence of the output power of a Xe−He laser for $\lambda = 3.50\mu$ on the diameter of the discharge tube, pressure of the gas mixture, pumping power, discharge length, etc., for the purpose of finding the optimum conditions under which the highest power output would be obtained.

For the experiment we used the laser described previously. We used plane mirrors (fused optical quartz backings), both with dielectric and with metallic coatings. The output power of the laser was measured in relative units.

The dependence of the Xe−He laser output power on the pressure in the discharge tube for λ equal to 3.50 and 3.36 μ and for tube diameters of 8, 12, and 20 mm is shown in Fig. 10. As the figure indicates, the maximum generated power is observed in the vicinity of pressures from 1.7 to 2 mm Hg. For λ = 3.36 μ and a discharge tube diameter of 8 mm, the maximum generated power occurs at p ~ 0.7 mm Hg. It is important to note that the laser has stable operation for tubes 8 and 12 mm in diameter over a wide range of gas pressures from 0.2 to 20 mm Hg or more. Generation is always stable and is observed both with continuous renewal of the gas and in a "sealed system," where there is no inclination for the gas mixture to become contaminated with the air.

The laser exhibits stable operation at a discharge length of 2 or 3 cm and a pumping power of 1 W.

Figure 11 shows the results of an investigation of the dependence of the laser output power on the pumping power for the 12-mm tube (λ = 3.5 and 3.36 μ). As the graph reveals, the maximum power is obtained for p ~ 2 mm Hg and P_{pump} ~ 50 W. At first the power rises steeply, then stays at one level as P_{pump} is increased. We were unable to increase P_{pump} any higher, because the mirrors could fail under the action of discharge.

All the investigations were conducted with a mixture of Xe and He with partial pressures in the respective ratio of 1 : 100.

It was established that the power of the laser increases linearly with the discharge length, beginning with L = 300 mm.

As the experimental investigations indicated, a Xe−He gas laser (λ = 3.50 μ) can also be fabricated with small dimensions and, recognizing the fact that λ = 3.50 μ is absorbed very little by the atmosphere, the laser described above should find broad practical applications.

5. Procedure for Measuring the Angular Divergence; Experimental Results

The spatial divergence of the laser beam is determined primarily by the influence of diffraction on the energy distribution between the various modes in the cavity and by the quality of the mirror surfaces [12].

The present experiment was set up with a view toward studying the beam divergence (λ = 1.15 μ) as a function of the laser output power. The method of measurement was based on angular measurements of the diffraction pattern obtained in the principal focal plane of the objective.

The observations were made as follows. An interference filter with transmission at the maximum λ = 1.15 μ was placed in the path of the laser beam. The beam was focused by means of an objective with a focal length f = 1600 mm onto photographic film. The output power was measured at the other end of the laser.

These measurements were conducted on a laser with a separation of 3 m between mirrors, a discharge length of 2 m, and a discharge tube with an inside diameter of 8 mm.

We first recorded the optical density curves of the film for λ = 1.15 μ (sensitivity of 1.2 units on the GOST scale). The curve was recorded, on the one hand, on a DFS-8 spectrograph by variation of the slit width and suitable choice of exposure and, on the other hand, using a laser beam with logarithmic attenuator. After this, the beam was photographed at various laser output powers.

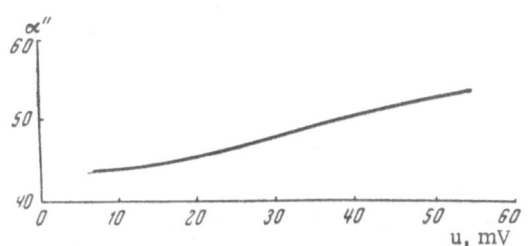

Fig. 12. Dependence of the angle of spatial divergence on the laser output power (power in relative units) for $\lambda = 1.15\,\mu$.

As we know, the diffraction pattern depends on the shape and dimensions of the apertures and the incident wavelength; the angular radius of the first dark ring is equal to

$$\alpha = 1.22 \cdot \frac{\lambda}{d},$$

where d is the diameter of the aperture.

The linear dimensions of the first dark ring $l = \alpha f$, where f is the focal length of the objective.

By measuring the dimensions of the spot obtained on the photographic film, the angle of divergence $\alpha = l/f$ was determined according to the above equation.

Before the photographing operation, the laser output beam was directed to the screen of an electron-optical converter, and the laser was carefully tuned in order to render the spot sharp and clear, and the power was measured to produce the maximum output for the given mode of operation, after which photographs were taken in the focal plane of the objective.

The dependence of the angle of spatial divergence on the output power of the laser in relative units is shown in Fig. 12. As apparent from the graph, the angle of divergence α is near the diffraction angle ($\alpha \sim 30''$) for small laser power values and increases slowly with the latter. The laser power was varied by varying the pumping power.

6. Investigation of the Modulation Characteristics of Lasers (Internal Modulation)

In view of the high monochromaticity and directionality of the radiation emitted by a laser, the latter may be used in communication systems for the transmission of large volumes of information ($f = 3 \cdot 10^{14}$ cps). There are two ways in which laser radiation can be modulated: 1) by internal modulation (modulation of the pumping energy); 2) external modulation of the output with a Kerr cell or crystals.

The objective of the experiment was to explore the possibility of modulating the radiation of lasers by the first technique. The experiment was conducted as follows. The radiation from the lasers was directed through an interference filter to the cathode of a photomultiplier, from which the signal was fed to the input of a DSO-1 double-beam oscilloscope. Discharge was excited in the tube by a high-frequency generator ($f = 28$ Mc), the frequency of which was amplitude-modulated by the frequency of a standard signal generator. The modulation of the laser output signal was measured by changing the frequency of modulation of the pumping generator.

As the experiment demonstrated, laser modulation by this technique can be realized without distortions up to 100 kc (with high-frequency discharge excitation).

We were able in practice to effect the transmission of audio signals (speech, music, telegraph signals) using an Ne−He gas laser ($\lambda = 1.15\,\mu$) under laboratory conditions. As shown by a theoretical calculation and experiment [17], currently available instrumentation is inadequate for precise spectral investigations of the radiation from a gas laser. We began, therefore, to seek a suitable procedure for such investigations. We arrived ultimately at the method of rotating lasers [18].

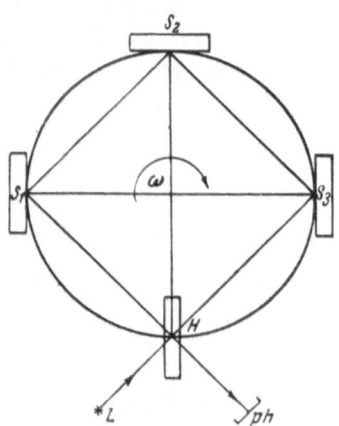

Fig. 13. Optical path of the beams in Sagnac's experiment.

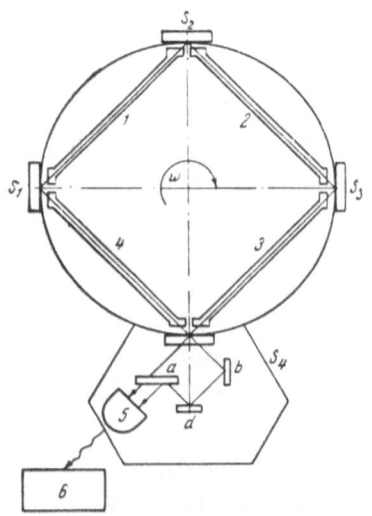

Fig. 14. Laser arrangement with an annular resonator.

7. Method of Measuring the Emission Spectrum of Lasers; Experimental Results

This experiment was based on the well-known experiment of Sagnac [19], which consists in the following. Three mirrors S_1, S_2, S_3 and a semitransparent plate H are placed at the corners of a square on a disc (Fig. 13). The disc can be rotated together with the source and photographic plate. The light from the source L is divided into two rays: One is transmitted through the plate H and is reflected from the mirrors in a counterclockwise direction, impinging on the photographic plate ph; the other is reflected from H and is transmitted in a clockwise direction, also impinging on ph. Interference takes place on the plate between two monochromatic rays. If the disc is rotated with an angular velocity ω, the ray whose direction of transmission coincides with the direction of rotation will impinge on ph with a certain delay relative to the ray whose direction of transmission is opposite the direction of rotation. The interference fringes that occur with the disc at rest undergo a certain shift on rotation of the disc, first in one, then in the other direction. The relative shift ΔZ of the fringes is measured, as expressed by the formula

$$\Delta Z = 4\beta \, \frac{S}{r\lambda},$$

where S is the area subtended by the light beam (area of the square), r is the radius of the disc, $\beta = v/c$, and v is the velocity of a point on the circumference. This equation may be written for the time differences:

$$\Delta\tau = 4\beta \, \frac{S}{rc}.$$

Sagnac confirmed the derived formula in practice.

The shift of the fringes ΔZ clearly may be obtained on the basis of the Doppler effect, which occurs here as a result of the quarter-wave plate H acting as a moving source, radiating different wavelengths forward and backward. We designed and built a device reproducing the experiment of Sagnac, but constructed with gas lasers as its base [17], which we then used to measure the emission spectrum of a laser. The arrangement is shown in Fig. 14. Four discharge tubes 8 mm in diameter, with $l = 50$ cm and windows at the Brewster angle, were mounted on a rotating platform. Three reflecting mirrors S_1, S_2, and S_3, and a semitransparent plate S_4 are placed at the corners of a square. The laser beam is transmitted through the semitransparent plate in mutually perpendicular directions, then is allowed to impinge on a photomultiplier by means of the mirrors a, b, and d (the traversal scheme followed by the beam can be realized using only one auxiliary mirror b set perpendicular to the beam).

On rotation of the platform the light beam propagating in the direction of rotation is delayed relative to the beam traveling counter to the direction of rotation. This produces a change in the phase difference of the two beams, which is accompanied by a modification of the beat spectrum on the photomultiplier. The degree of monochromaticity of the laser radiation can be determined from the difference frequency of the beats.

Work is currently in progress along the following lines: 1) measurement of the intrinsic linewidth of a gas laser due to spontaneous noise; 2) measurement of the output emission spectrum of a laser as a function of its power; 3) investigation of the statistical characteristics of the output emission; and, 4) measurement of the technical linewidth of a laser due to instability of the pumping source, ambient temperature fluctuations, and other factors.

CHAPTER III

EFFECT OF THE DISCHARGE TUBE GEOMETRY
ON THE KINETIC PROCESSES OF A GAS LASER

It was shown in Chapter II that the output power of a gas layer is critical with respect to the characteristic dimensions of the discharge tube. We now propose to analyze the influence of the tube geometry on the kinetic processes in the plasma of a neon−helium laser. Although the calculations are carried out for the specific transition† $2p^54s^1P_1 - 2p^53p^3P_2(2s_2 - 2p_4)$ (Fig. 15), the processes leading to the formation of an inverted population are fairly general and may be applied to a number of other transitions in the discharge plasma of a gas laser.

1. Statement of the Problem

In the pure helium gas laser the generated power is relatively low. The addition of helium to the neon increases the power output immensely by virtue of an increase in the inverted population and makes it possible to realize generation in new transitions. We summarize briefly the factors leading up to the enhanced population of working levels.

1. An increase in the electron concentration [20, 21]. A possible reason for this is the formation of molecular ions in correspondence with the reaction‡ $He^* + He \rightarrow He_2^+ + e$.

2. An increase in the average energy of the electrons [20, 21]. One process that could eventually lead to increased energy might be the decay of molecular ions [22-23], which, like diffusion, is associated with the absorption of electrons and ions. The electron energy might also be similarly affected by the presence of electrically negative impurities (such as water vapor, for instance).

3. The transfer of impact excitation of the second kind from excited helium to normal neon atoms. This case is realized, for example, when the $2p^54s^1P_1(2s_2)$ level of neon becomes populated [1, 24, 25].

4. Selective population of levels by the decay of molecular ions [22-23]. An excited neon or helium atom, on colliding with a normal atom, forms a molecular ion of the type He^+Ne, $HeNe^+$, He_2^+, Ne_2^+, etc. The collision of the latter with a slow electron results in dissociative recombination with the formation of two neutral atoms, one of which is excited. Dissociative recombination leads to selective population of the neon levels.

The processes 1-2, which are related to an increase in the temperature and concentration of electrons, has a rather weak effect on the generation of the majority of transitions [20-21].

---

† Here and elsewhere the symbols in parentheses refer to the Paschen series notation.

‡ The asterisk(*) denotes the excited state of the atom.

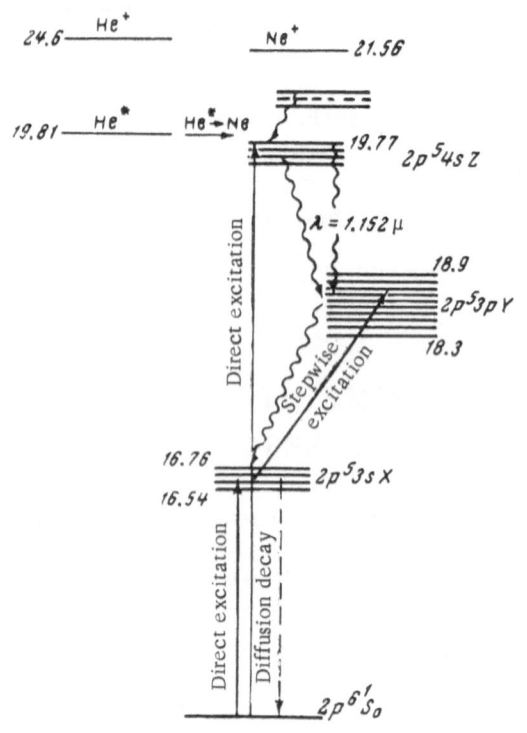

Fig. 15. Energy level diagram of Ne and He.

Dissociative decay and the resonance transfer of excitation play a far more important role.

In the discussion that follows we consider laser generation at the transition $2p^54s^1P_1 - 2p^53p^3P_2(2s_2 - 2p_4)$. All of the processes indicated above contribute to the population of the working levels in different degrees. We will examine the role of one of them, resonance transfer between helium and neon. Also of indisputable interest is the inclusion of dissociative recombination, which is important in the population of a number of gas levels. However, in the excitation of the level $2p^54s^1P_1$, which is close to the helium metastable level $1s2s^3S_1$, the process He* → Ne is the most significant. It is also conceivable that the cross section obtained experimentally for this process, $\sigma \simeq 3.7 \cdot 10^{-17}$ cm^2 [27], takes dissociative recombination into account simultaneously.

2. Density of Electrons and Metastable Atoms of Neon and Helium

In the plasma of a gas laser there are a number of microprocesses that occur, affecting the buildup of the equilibrium concentration of electrons in varying degrees. As estimates have shown, the main source of electrons is the electronic ionization of atoms. The electron influx elicited by ionization is balanced out by diffusion and recombination with atomic and molecular ions. For characteristic discharge tube dimensions $\Lambda \sim 1$ cm and a gas pressure $p \sim 1$ mm Hg† recombination absorption has a relatively weak effect on the steady-state value of the electron density. The disappearance of electrons is primarily attributable to their diffusion toward the walls of the discharge tube. We will confine our ensuing discussion exclusively to this mechanism of electron losses.

At a high concentration of metastable neon and helium atoms the indicated microprocesses are accompanied by still another, stepwise ionization of the gas. With the latter taken into account, the time variation of the electron density n is described by the equation

$$\frac{\partial n}{\partial t} = \sum n \langle \sigma_{0i} c \rangle N_{i0} + \sum n \langle \sigma_{*i} c \rangle N_i^* + D_j \nabla^2 n. \qquad (1)$$

Here N_{i0}, σ_{0i}, and N_i^*, σ_{*i} are the densities and effective ionization cross sections for the ground and metastable states of helium (i = 1) and neon (i = 2), respectively; c is the electron velocity; D_j is the ambipolar diffusion constant, and ∇ is the Laplace operator. The averaging of $\langle \sigma c \rangle$ is carried out over the velocity distribution of the electrons, the summation over the components of the gas.

In the discharge of a gas laser comparatively few atoms are found in the excited state. Consequently, the spatial distribution of electrons is adequately described by the zeroth-order

_____
† Such conditions are customary for a discharge in a gas laser.

Bessel function† $J_0(2.4r/a)$. In other words, we assume that the nonlinear terms of Eq. (1) are expressed only in the "amplitude" of the electron density, not in their spatial distribution.

This approximation is also applicable to the calculation of the density of metastable atoms. Although metastable atoms are not totally disintegrated due to collisions with the tube walls, we will postulate that their spatial distribution copies the behavior of the electron density. Therefore, replacing n(r) by $n_0 J_0(2.4r/a)$ and N_i^* by $N_{i0}^* J_0(2.4r/a)$ and integrating (1) over the entire tube volume (in the steady state $\partial n/\partial t = 0$), we obtain

$$\sum n_0 \langle \sigma_{0i} c \rangle N_{i0} + \sum n_0 \langle \sigma_{*i} c \rangle N_{i0}^{\cdot} \int J_0^2 dV / \int J_0 dV - D_i n_0 S |\nabla J_0| / \int J_0 dV = 0. \tag{2}$$

Here S is the surface area of the tube and is approximately equal to $2\pi a l$; the derivative ∇J_0 is evaluated at $r = a$.

The population of the metastable helium level $1s2s^3S_1$ is determined from the expression

$$N_{10}^{\cdot} = \frac{n_0 N_{10} (\langle \sigma_{01} c \rangle + \sum \langle \sigma_{01}^{\cdot} c \rangle)}{SD |\nabla J_0| / \int J_0 dV + \sum N_{i0} \langle \sigma_{12} v_i \rangle + n_0 \langle \sigma_{*i} c \rangle \int J_0^2 dV / \int J_0 dV}, \tag{3}$$

where N_{10} and N_{20} are the densities of the normal helium and neon atoms; $\langle \sigma_{01} c \rangle$ and $\langle \sigma_{01}^* c \rangle$ are the occupation rates of the level $1s2s^3S_1$ by electronic excitation of the ground state and radiative decay of higher levels, respectively; D is the diffusion coefficient $\sum N_{0i} \langle \sigma_{12} v_i \rangle$; and $n_0 \langle \sigma_{*i} c \rangle$ are the decay rates of metastable helium atoms due to collisions with normal helium and neon atoms [24, 27] and electronic ionization, respectively.

For small discharge tube diameters and electron densities $n_0 \sim 10^{11}$ diffusion disintegration plays the most important role in the depopulation of metastable atoms, where the rate of this process $SD |\nabla J_0| / \int J_0 dV$ is inversely proportional to the tube radius squared. Making use of the numerical expression for $SD |\nabla J_0| / \int J_0 dV + \sum N_0 \langle \sigma_{12} v_i \rangle$ [27], and recognizing the resonance character of the process He* → Ne, we obtain the following for the decay rate of the level $1s2s^3S_1$ (disregarding electronic depopulation)

$$\frac{1}{\tau^{\cdot}} = \frac{5.8}{a^2} \frac{\left[3kT_i \left(\frac{1}{M} + \frac{1}{M^*}\right)\right]^{1/2}}{3\pi (\rho + \rho^*)^2 \sum N_{i0}} + 2\mu (\rho + \rho^*)^2 \left[2\pi kT_i \left(\frac{1}{M} + \frac{1}{M^*}\right)\right]^{1/2} \sum N_{i0} + N_{20} 3.7 \cdot 10^{-17} v_i. \tag{4}$$

Here T_i is the gas temperature, M* and ρ^* are the mass and radius of the metastable helium atom, M and ρ are the mass and radius of the atom with which the metastable helium atom collides, and μ is the probability of disintegration of metastable atoms in such a collision.

In the population of metastable neon, the resonance transition He* → Ne is of considerable importance in addition to direct electronic excitation and radiative decay. Metastable neon atoms populate four $2p^5 3s(X)$ levels, two of which ($2p^5 3s^1 P_0$ and $2p^5 3s^3 P_1$) are optically coupled with the ground state. For neon pressures typical in a laser (p ~ 0.1 mm Hg) the resonance absorption of phonons radiated in the decay of the states $2p^5 3s^1 P_1$ and $2p^5 3s^3 P_0$ is so effective that all four 2p3s levels may be regarded as metastable.

Consequently, in the steady state the density of metastable neon atoms satisfies the

---

† We are assuming that the discharge tube is cylindrical, with radius a and length $l \gg a$.

equation

$$N_{20}^* = \frac{n_0 N_{20} \left(\sum \langle \sigma_{02}c \rangle + \sum \langle \sigma_{02}^*c \rangle \right) + N_{10}^* N_{20} \langle \sigma_{12}v_i \rangle}{SD |\nabla J_0| / \int J_0 dV + \sum N_{i0} \langle \sigma_{12}v_i \rangle + n_0 \sum \langle \sigma_{*i}c \rangle \int J_0^2 dV / \int J_0 dV}. \tag{5}$$

Here $n_0 \Sigma \langle \sigma_{02}c \rangle$ and $n_0 \Sigma \langle \sigma_{02}^*c \rangle$ are the rates of occupation of metastable levels by electronic excitation and decay of higher levels, $N_{10}^* \langle \sigma_{12}v_i \rangle$ is the rate of the resonance transition He* → Ne [24], and $SD |\nabla J_0| / \int J_0 dV$, $\Sigma N_{i0} \langle \sigma_{12}v_i \rangle$, $n_0 \Sigma \langle \sigma_{*i}c \rangle$ are the rates of diffusion decay, disintegration by collisions with gas atoms, and ionization, respectively.

The quantity N_{20}^* represents the total density of the four $2p^5 3s$ levels. Due to collision with atoms, these levels are intermingled, and since the energy intervals separating them are small, the population of all four levels may be regarded as essentially the same. This is qualitatively confirmed by the experimental data [28].

Now the system of equations (2)-(5) places us in a position to find the density of electrons and metastable neon and helium atoms as a function of the electron temperature of discharge,† which is completely determined by the partial pressure of the gases, the radius of the discharge tube, and the pumping power.

The electron temperature T_e may be found on the basis of the energy balance:

$$n_0 \int_0^\infty P_x(c) \left(\frac{m}{2\pi k T_e} \right)^{3/2} e^{-\frac{mc^2}{2kT_e}} 4\pi c^2 dc = \frac{Pl}{\int J_0 dV}. \tag{6}$$

Here $P_x(c)$ is the power dissipated by the electron in elastic and inelastic collisions with gas atoms and P is the power pumped into the discharge tube per unit length. Taking into account the energy losses of the electron, according to [30, 32], it is possible to derive an expression linking the temperature with the pumping power and diameter of the discharge tube.

3. Populations of the Working Levels of Neon

If the intensity of the field radiated by the atoms is denoted by E, the Einstein coefficient of the induced transition $2p^5 4s^1 P_1 - 2p^5 3p^3 P_2 (2s_2 - 2p_4)$ by w, the population of the upper working level $N(2s_2)$ is satisfied by the equation

$$N_{10}^* N_{20} \langle \sigma_{12}v_i \rangle + N_{20}n_0 \langle \sigma_{02s}c \rangle = \left\{ wE^2 [N(2s_2) - N(2p_4)] + \frac{1}{\tau_{2s}} N(2s_2) \right\} \frac{V}{\int J_0 dV}. \tag{7}$$

The left-hand side of the above equation describes the occupation of the upper working level by the resonance transition He* → Ne and electronic excitation (with regard for cascade decay), the right-hand side describes induced and radiative decay of the level $2p^5 4s^1 P_1$.

The population of the lower working level $N(2p_4)$ satisfies the relation

$$wE^2 [N(2s_2) - N(2p_4)] V + \frac{1}{4} N_{20}n_0 \sum \langle \sigma_{sp}c \rangle \int J_0^2 dV + N_{20}n_0 \langle \sigma_{02p}c \rangle \int J_0^2 dV = N(2p_4) \frac{1}{\tau_{2p}} V. \tag{8}$$

---

† In any discharge similar to that in a neon−helium laser, the energy distribution of the electrons will clearly depart from a Maxwellian law [29]. However, we will assume for computational simplicity that the electron energy distribution is Maxwellian. This is not reflected qualitatively in any of the processes with which we are concerned herein.

Here $\Sigma <\sigma_{sp}c>$ is the total probability of excitation of the level $2p^53p^3P_2$ (for the allowed transitions $2p^53p^3P_2 - 2p^53s$), $<\sigma_{02p}c>$ is the probability of radiative population by the decay of higher levels, and $1/\tau_{2p}$ is the probability of radiative decay of the level $2p^53p^3P_2$.

Equations (7) and (8) do not form a closed system, insofar as they do not account for the influence of the resonator Q on the rate of induced decay of the working levels. Recognizing that the Q of the resonator is related to the induced radiation power P_{out} and the energy W stored in the resonator by the expression $P_{out}/W = 2\pi\nu/Q$, and replacing P_{out} by $[N(2s_2) - N(2p_4)]wE^2\hbar\nu V$ and W by $E^2V/4\pi$, we obtain for the inverted population

$$N(2s_2) - N(2p_4) = 1/4\pi Q\hbar w. \tag{9}$$

Then, replacing the coefficient of induced emission w and the Q by the respective quantities $c_0^3/\tau_{21}8\pi^3\nu^2\Delta\nu$ and $2\pi L/\lambda (1-\alpha)$ [where $1/\tau_{21}$ is the probability of the spontaneous transition $2p^54s^1P_1 - 2p^53p^3P_2 (2s_2 - 2p_4)$, $\Delta\nu$ is the Doppler linewidth, ν is the frequency of the working transition, c_0 is the velocity of light, α is the transmission and diffraction loss coefficient, and L is the length of the resonator], we obtain an equation closing the system (7), (8).

Now it is possible in general form to find the coherent radiation field as a function of the geometric dimensions of the discharge tube and pumping power. All that is needed for a numerical calculation are the specific values of the effective cross sections for electronic excitation of the levels.

4. Electronic Excitation Cross Section

For an estimate of the inverted population of the working levels it is appropriate to use (for electron energies of 3-5 eV) various approximations of the average effective cross sections. The probability of electronic excitation of the ground state may be found in accordance with [31, 32].

Adopting, for example, the following model of the effective cross section [32]:

$$\sigma = \begin{cases} \sigma_0 \dfrac{(c^2 - c_e^2)^{3/2}}{c_e c^2} & \text{for } c > c_e, \\ 0 & \text{for } c < c_e, \end{cases}$$

we obtain

$$\langle \sigma c \rangle = \frac{3\sigma_0}{c_e}\frac{kT_e}{m} e^{-\frac{\mathcal{E}_e}{kT_e}}. \tag{10}$$

For the excitation of a metastable level the approximation (10) is unsuitable, because the average electron energy is comparable with the extremal energy for the excitation cross section. In this case the probability of excitation may be found according to [33]. At electron temperatures near the energy spacings of the metastable and excited levels (3-5 eV) the approximation of $<\sigma c>$ according to [33] yields results nearly equal to

$$\langle \sigma c \rangle \simeq \sigma_m \left(\frac{8}{\pi}\frac{kT_e}{m}\right)^{1/2}. \tag{11}$$

Here σ_m is the maximum value of the cross section.

This is the approximation we will make use of below. Expressions analogous to (10) and (11) are obtained also for the electronic ionization of an atom with ground and metastable states.

The maximum cross sections for ionization from the ground levels of neon and helium are equal to $0.85 \cdot 10^{-16}$ and $0.63 \cdot 10^{-16}$ cm^2, respectively [32]. The cross sections for ionization of metastable atoms may be estimated with an error given by the semi-empirical formula [34]:

$$\sigma(\mathscr{E}) = 2.66 \xi \pi a_0^2 \left(\frac{\mathscr{E}_h}{\mathscr{E}_i}\right)^{1/2} \frac{\mathscr{E}/\mathscr{E}_i - 1}{(\mathscr{E}/\mathscr{E}_i)^2} \log 1.25 \cdot \frac{\mathscr{E}}{\mathscr{E}_i}. \tag{12}$$

Here $\mathscr{E}$ is the electron energy; $\mathscr{E}_i$ is the ionization energy. The parameters $\xi = 2$ for ionization from the ground state, $\xi = 1$ for the ionization of metastable atoms. According to (12), the ratio of the ground level ionization cross section to the metastable level ionization cross section is equal to

$$\sigma_m / \sigma_m^{\bullet} = [\sqrt{\overline{\xi} \mathscr{E}_i^{\bullet}} / \sqrt{\overline{\xi^{\bullet}} \mathscr{E}_i}]^{1/2}. \tag{13}$$

Hence for the maximum ionization cross sections of metastable neon and helium atoms we obtain values of $8 \cdot 10^{-16}$ and $4.9 \cdot 10^{-16}$ cm^2, respectively. These values are an order of magnitude greater than the direct ionization cross section.

Data regarding the electronic excitation of the lower working level ($2p^5 3p\, ^3P_2$) and the metastable levels ($2p^5 3s\, ^1P_1$ and $2p^5 3s\, ^3P_2$) of neon from the ground state have been obtained in [35-37].

Metastable neon is not populated only by direct excitation of the ground state and the resonance transition He* → Ne. Radiation decay of the higher levels also plays an important part. Utilizing the numerical data of [37], we estimate the cross sections for the cascade population of metastable neon. It is approximately one order of magnitude larger than the cross section for electronic excitation of the level $2p^5 3p\, ^3P_2$.

The effective cross sections of the transitions $2p^5 3p\, ^3P_2 - 2p^5 3s$ may be found in [28, 37]. The cross sections of the transition $2p^6\, ^1S_0 - 2p^5 4s\, ^1P_1$, $\sigma \simeq 5 \cdot 10^{-19}$ cm^2, have been found in the Born approximation by a computer program [38]. The rates of radiative decay of the levels participating in the creation of the inverted population of the transition $2p^5 4s\, ^1P_1 - 2p^5 3p\, ^3P_2$ are taken from [24], data on the electronic excitation of helium from [39-40].

5. Laser Output Power

Equations (7)-(9), taking into account the effective cross sections of electronic and radiative excitation of the levels, make it possible to estimate the output power of the neon−helium laser [42]. The calculations presented below were carried out on the M-20 computer.

The dependence of the output power on the diameter of the discharge tube at neon and helium partial pressures of 0.1 and 1 mm Hg, respectively, and a resonator length $L = 2 \cdot 10^2$ cm is shown in Fig. 16. As the diameter of the tube is increased, the output power rises at first, reaching a maximum at 6 mm, then falls off sharply. For large tube diameters (25-30 mm) generation essentially does not occur.

The existence of an optimum tube diameter is attributable to the counteraction of two effects. On the one hand, as the diameter is decreased, the volume of working gas and the power associated with it diminish. On the other hand, with decreasing diameter the inverted population increases (for large electron densities, at any rate) and, as a result, the power radiated in the generation of unit volume increases.

The increase in the inverted population with diminishing diameter is due first and foremost to the increasing energy of the electrons. In fact, the average electron energy is not determined principally by the pumping power, but solely by the geometry of the discharge tube. With decreasing diameter the diffusion leakage of electrons at the walls increases. This results

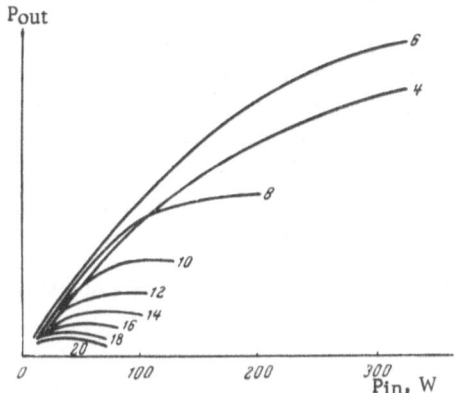

Fig. 16. Dependence of the output power P_{out} on the pumping power P_{in} for various discharge tube diameters. Tube length l = 2 m, transmission coefficient of the mirrors 2%. P_{out} is given in relative units. The tube diameter is indicated in mm next to each curve.

in an increased electron temperature. Disregarding stepwise ionization, the electron temperature satisfies the equation

$$\gamma_i n + D_j \nabla^2 n = 0.$$

Here γ_i is the ionization frequency and D_j is the ambipolar diffusion constant. Replacing γ_i in accordance with (10), and expressing D_j in terms of the ion mobility b_j and electron temperature, we obtain

$$\sum N_{io} \frac{3\sigma_{oi}}{c_i} \frac{kT_e}{m} e^{-\frac{\mathscr{E}_i}{kT_e}} = \frac{b_i}{e} kT_e \frac{\int \nabla n(r) d\vec{S}}{\int n(r) dV}.$$

If we introduce the characteristic dimension Λ of the discharge tube ($\Lambda^2 \sim |\int n(r)dv / \int \nabla n(r)dS|$), the equation for the temperature is written

$$\sum N_{io} \frac{3\sigma_{oi}}{c_i} e^{-\frac{\mathscr{E}_i}{kT_e}} \sim \frac{b_j}{\Lambda^2}. \tag{13a}$$

For a cylindrical tube with radius a much greater than its length, $\Lambda^2 = a^2 / 5.8$. Consequently, a reduction in the radius produces an increase in the electron temperature, and the population rate of the levels grows as a^{-2}.

The inverted population does not depend only on the electron temperature. The temperature dictates its upper boundary, the nearness to which is also governed by the electron density. At low pumping powers (small electron densities and the interval of linear growth of the output power) the electron density in a tube of diameter d_1 can exceed the electron density in a tube of diameter d_2 (where $d_2 < d_1$) to such an extent, despite the higher electron temperature in the latter, that the rate of population of the working levels in the former will be higher. This also explains the intersection of the curves in Fig. 16. For large electron concentrations the population reaches its maximum, and a further increase in the pumping power only yields (as the result of electronic depopulation of the excited states and predominant filling of the lower working level†) a reduction in the output power (see Fig. 15). This is particularly manifested in the case of generation in tubes of large radius; even with small pumping powers in such event, the electron densities are quite high.

A comparison of Fig. 16 with Fig. 5 indicates qualitatively the same behavior on the part of the theoretical and experimental curves and satisfactory agreement between the optimum diameters of the discharge tubes; the calculated diameter‡ is equal to 6 mm, the experimental value to 8 mm. Replacement of the helium-neon resonance transition process by dissociative recombination in the populating of the level $2p^54s^1P_1(2s_2)$ (recognizing that dissociative recombination plays the predominant part in the excitation of this level) does not alter the value of the

---

† The population of the upper working level is proportional to n, that of the lower level to n^2.

‡ The optimum diameter may be found for varying pressure from the relation $pd = p_0 d_0$, where d_0 is the optimum diameter for $p = p_0$. But since the inverted population cannot, in general, be represented as a function of pd, this relation is actually more qualitative than quantitative.

optimum diameter appreciably. This is a consequence of the fact that the optimum diameter is determined (in the first approximation) by the ratio of the population of the upper and lower working levels, while the population of the lower level in turn is related in large part to the decay of the level $2p^54s^1P_1$. For this reason, the optimum diameter, clearly, will not be critical with respect to certain errors in determining the excitation cross sections of the levels or to the form of the electron energy distribution function. This is not true, of course, as regards the magnitude of the output power, which depends on the absolute value of the populations of the levels $2p^54s^1P_1$ and $2p^53p^3P_2$, not on their ratio.

We now list briefly the ways in which it might be possible to enhance the power output of the neon−helium laser.

1. Addition of a buffer gas [25] to intensify the decay of metastable neon atoms.

2. Increase in the temperature T_i of the working gas, thereby increasing the electron energy.

The increased average energy of the electrons is related to the accelerated diffusion of electrons and ions toward the tube walls. As a result, the frequency of electronic excitation of the ground levels increases in proportion to† $\sqrt{T_i}$; the population of metastable neon and helium atoms remains unchanged (for moderate electron densities, at any rate). At the same time, the frequency of collisions of the second kind $He^* \rightarrow Ne$ grows as $\sqrt{T_i}$. This causes an increase in the population of the upper working level. The population of the lower level, which is filled primarily by electronic excitation of metastable neon, remains practically unchanged. Consequently, when the excitation threshold is high enough, the power generated by each mode increases as $\sqrt{T_i}$. However, such arguments may prove invalid for other mechanisms of filling the levels.

3. Impulse excitation, which in the period after illumination results in predominant population of the upper working level [25].

4. Increase in the length of the resonator L. In this case, a large number of modes fall within the Doppler linewidth. If the frequency spacing of the modes exceeds the intrinsic linewidth of the radiation and the diffraction losses are small in comparison with the mirror transmission losses, the total power will increase in proportion to the resonator length.

5. Increase in the length of the discharge tube l, inducing a proportionate increase in the output power,

6. Increase in the characteristic dimension of the discharge tube Λ while holding the volume of the working gas constant or increasing it. The power generated per unit volume in this case grows at least as rapidly as Λ^{-2}.

This dependence of the power on Λ is related to the electron temperature and must therefore be realized for all transitions in the gas-discharge laser.

The indicated requirements for Λ are easily satisfied for a discharge tube with a rectangular cross section. If one side of the cross section is much shorter than the other, the characteristic dimension of the discharge tube will be determined by the smaller side. Here, as in the case of the circular tube, there will be an optimum value for the smaller side: $h = h_{opt}$. But now h_{opt} is not determined by the variation of the generated power with the changing volume of the working gas, but by the increase in the diffraction losses‡ with diminishing h. Clearly,

---

† Due to the mobility of the ions b_i increasing in proportion to their velocities [see Eq. (13)].

‡ The diffraction losses increase with diminishing h as h^{-3} [41], whereas the population of the levels does not grow any more rapidly than h^{-2}.

for $h \sim h_{opt}$ the diffraction losses become at least comparable with the mirror transmission losses. For $\lambda = 1 \mu$ and a resonator length $L = 10^2$ cm, this yields $h \sim 10^{-2}$ cm. Consequently, the effect associated with the optimum dimensions of the rectangular tube, given realistic resonator dimensions ($L \sim 10^2$ cm), is pronounced only in the infrared region in practice. In the visible or ultraviolet region of the spectrum a reduction in h should only increase the output power.

The authors close with utmost thanks to Yu. P. Trokhin, V. N. Lukanin, B. I. Prokopov, B. I. Belov, F. S. Titov, and A. F. Suchkov for their assistance in carrying out the investigation, as well as for a discussion of the results and sharing in the computations.

LITERATURE CITED

1. A. Javan, W. Bennett, and D. Herriot, Phys. Rev. Letters, 6 : 106 (1961).
2. S. Tolansky, High-Resolution Spectroscopy, p. 363 [Russian translation], IL, Moscow (1955) [English edition: Methuen (1946)].
3. H. D. Polster, J. Opt. Soc. Am., 42 : 21 (1952).
4. W. W. Rigrod, H. Kogelnik, D. J. Brangaccio, and D. R. Herriot, J. Appl. Phys., 33 : 743 (1962).
5. A. G. Cox and T. Li, Bell System. Tech. J., 40 : 453 (1961).
6. R. M. Baker, Electronics, 26(5) : 36-38 (1963).
7. V. Zuev and P. G. Kryukov, Pribory i Tekhn. Eksperim., No. 3, pp. 188-189 (1963).
8. A. Javan, Phys. Rev. Letters, 3 : 87 (1959).
9. N. G. Basov, E. P. Markin, and D. I. Mash, Zh. Eksperim. i Teor. Fiz., 43 : 1116 (1962).
10. D. Herriot, J. Opt. Soc. Am., 52(1) : 31-37 (1962).
11. N. G. Basov, E. P. Markin, and V. V. Nikitin, Opt. i Spektroskopiya, 15 : 436 (1963).
12. W. R. Bennett, Jr., Gaseous Optical Masers, Appl. Opt., Suppl. No. 1 to Optical Masers, 1962, pp. 24-62.
13. N. G. Basov, E. P. Markin, and V. V. Nikitin, Radiotekhn. i Elektron., 8 : 2084 (1963).
14. W. L. Faust, R. A. McFarlane, C.K.N. Patel, and C. G. Garrett, Appl. Phys. Letters, 4(1) : 85 (1962).
15. R. A. Paananen and D. L. Bobroff, Appl. Phys. Letters, 2(5) : 99-110 (1961).
16. E. P. Markin and V. V. Nikitin, Opt. i Spektroskopiya, 17 : 953 (1963).
17. A. Javan, Third International Conference on Quantum Electronics, Paris-New York (1963).
18. Electronic News, No. 359, pp. 1-27 (1963).
19. G. Sagnac, Compt. Rend., 157 : 708 (1913); A. A. Sommerfeld, Optics, p. 111 [Russian translation], IL, Moscow (1953).
20. A. Aisenber, Appl. Phys. Letters, 2(2) : 187 (1963).
21. H. Statz, C. L. Tang, and G. F. Koster, J. Appl. Phys., 31 : 2625 (1963).
22. M. P. Bogdanov, O. P. Bochkova, and S. É. Frish, Dokl. Akad. Nauk SSSR, 156(1) : 54 (1964).
23. V. S. Egorov, Yu. G. Kozlov, and A. M. Shukhtin, Opt. i Spektroskopiya, 15 : 839 (1963).
24. W. R. Bennett, Advances in Quantum Electronics. Proc. of the Second Internat. Conference, Columbia University Press, New York (1960).
25. W. R. Bennett, Usp. Fiz. Nauk, 81 : 119 (1963).
26. Yu. M. Kagan and R. I. Lyagushenko, Zh. Tekhn. Fiz., 32 : 193 (1962).
27. A. Éngel' and M. Shtenbek, Physics and Engineering of Electrical Discharge, Vol. 1, NTI NKTP SSSR, Moscow (1935).
28. Yu. M. Kagan, R. I. Lyagushenko, and A. P. Khakhaev, Opt. i Spektroskopiya, 14 : 593 (1963).
29. N. A. Vorob'eva, Yu. M. Kagan, and V. M. Millenin, Zh. Tekhn. Fiz., 34 : 828 (1964).
30. V. L. Granovskii, Electric Current in a Gas, Vol. 1. GITTL, Leningrad (1952).

31.   S. É. Frish and V. E. Yakhontova, Opt. i Spektroskopiya, 4:402 (1958).

32.   T. Kikhara, Rev. Mod. Phys., 24:45 (1962).

33.   I. I. Sobel'man, Introduction to the Theory of Atomic Spectra. Fizmatgiz, Moscow (1963).

34.   H. W. Drawin, Z. Phys., 164:513 (1961).

35.   A. Gold and R. S. Knox, Phys. Rev., 113:834 (1959).

36.   F. A. Korolev, V. I. Odintsova, and E. V. Fursova, Opt. i Spektroskopiya, 16:555 (1964).

37.   S. É. Frish and V. F. Reval'd, Opt. i Spektroskopiya, 15:726 (1963).

38.   L. A. Vainshtein, Tr. Fiz. Inst. Akad. Nauk SSSR, 15:3 (1959).

39.   A. Eucken and K. H. Hellwege, Landolt-Börnstein Atom- und Molekularphysik, 1:341 (1950).

40.   G. F. Drukarev, Theory of Electron Collisions with Atoms, p. 183. Fizmatgiz, Moscow (1963).

41.   L. A. Vainshtein, Zh. Eksperim. i Teor. Fiz., 44:1050 (1963).

42.   É. M. Belenov and A. I. Oraevskii, Opt. i Spektroskopiya, 18:858 (1965).

DESIGN PROBLEMS AND INVESTIGATION OF THE
PERFORMANCE OF THE HYDROGEN ATOMIC
BEAM QUANTUM OSCILLATOR

N. G. Basov, G. M. Strakhovskii, A. I. Nikitin,
T. F. Nikitina, V. M. Tatarenkov, and A. V. Uspenskii

INTRODUCTION

The last decade has witnessed the vigorous development of a whole new branch of science, quantum electronics. The breakthrough that gave quantum electronics its powerful impetus forward was the development of a molecular oscillator utilizing a beam of ammonia molecules [1-3]. The progress of quantum electronics has made it possible to generate high-stability frequency standards. The basic factor limiting the stability of present-day frequency standards is the short interaction time between molecules or atoms and the high-frequency field, i.e., excessive linewidth. Several methods have been proposed for enhancing the time of interaction between the atom and the high-frequency field, for example, the method of separated oscillatory fields [4], which is essentially based on an increase in the equivalent pathlength of the atom in the high-frequency field; another tentative method involves the use of a beam of slow atoms or molecules [5].

In 1956, Ramsey indicated the conceptual possibility of improving the method of separated oscillatory fields [6]. He suggested increasing the time of interaction between the atom and high-frequency field by interposing a storage box in the path of the beam, in the gap between two separated resonators. This increased the incumbency time of the beam atoms in the gap between the two resonators due to their collision with buffer gas atoms or with the box walls. An inevitable condition for the realization of this notion is the requirement that the atom not change its quantum state in collision, thus limiting the method, as first pointed out by Dicke [7], strictly to atoms in the ground state. The best working substance from the point of view of this requirement is atomic hydrogen, which has a very low polarizability [8]. However, hydrogen atoms are difficult to detect, hence the first arrangement utilizing the principle of stored atoms was constructed using a beam of cesium atoms [8, 9].

Teflon was tried as the first coating for the walls of the storage box. The authors subsequently tested more than ten substances [8], with a particular leaning toward dimethyl dichlorsilane, on which Dicke and Wittke [10] also concentrated their attention. Of all the substances tested, however, the most suitable for cesium atoms proved to be a coating made up of a mixture of heavy paraffin. It was possible, using a box with this type of coating, to reduce the resonance linewidth to 150 cps, corresponding to approximately 60 collisions of cesium atoms with the walls. A highly anticipated result of comparing the polarizabilities of hydrogen and cesium atoms was that hydrogen atoms would suffer about 10^5 collisions with paraffin-coated walls. This many collisions would be tantamount to the atom lingering in the high-frequency field for a period on the order of one second, thus offering the possibility of performing spectroscopic measurements with astounding accuracy. On the other hand, the feasibility of obtaining such a long interaction time stimulated the notion of trying to use the emission of hydrogen atoms at the hyperfine transition frequency for the creation of a quantum oscillator. As a matter of fact, the magnetic dipole moment of the transition between levels of the hyperfine structure is approximately $1/100$ of the electric dipole moment (of, say, the ammonia molecule [2]). The oscillator self-excitation condition involves the square of the dipole moment and the square of the interaction time between the particles and the high-frequency field; the smallness of the dipole moment can offset the increased interaction time. It is necessary in order for the self-excitation conditions to be realized that the average interaction time between the atom and the field be about one second, which is indeed realizable in principle.

Goldenberg, Kleppner, and Ramsey built a quantum oscillator with a beam of hydrogen atoms. In 1960, they succeeded in producing oscillation at the 1420.4-Mc hyperfine transition frequency of a hydrogen atom in the ground S-state [11, 12]. The source of atomic hydrogen was a Wood discharge tube, the storage vessel was made of fused quartz and was coated on the inside with paraffin (in the first experiments) or dimethyl dichlorsilane (in later experiments). For these coatings, the lifetime of the atoms was at least 0.3 sec (about 10,000 collisions of atoms with the wall with no change of state). The resonance linewidth was several cycles per second, corresponding to a line quality $Q_l \sim 10^9$.

The first-order Doppler effect strongly inhibits the procurement of narrow lines in radiospectroscopy (for example, in a radiospectroscope with a gas cell [13]). The use of molecular beams with their velocity vector perpendicular to the direction of electromagnetic wave propagation made possible narrower absorption lines [14]. In ammonia molecular oscillators the same principle was used for suppression of the Doppler shift; wave modes and oscillator dimensions were chosen such that the phase velocity of the propagating wave would be almost infinite in the direction of motion of the beam [1]. For the atoms in the storage vessel there is no distinct direction of the velocity, and we come right back again to the gas cell with all its inherent shortcomings. But, as indicated in [7, 10], the first-order Doppler effect can be almost completely overcome if the dimensions of the storage box are smaller than the emission wavelength of the atoms. This fact has largely contributed to the success of efforts aimed at building a hydrogen oscillator. After the debut of the first hydrogen oscillator in the USA in 1960, oscillators were designed and put into operation in Switzerland (in 1962) [15] and the USSR (in 1963). The theory of the operation of the hydrogen quantum oscillator is presented in [16]. The theory is formulated analogously to the theory of the ammonia molecular oscillator [2, 3, 17-19]. The construction of the device is described in [20-23].

The over-all dimensions of the devices were made smaller than those of the first prototype, and teflon was recognized as the best coating for the storage vessel.

After the assurance of a sufficiently high stability on the part of the oscillation frequency, a precise measurement was made of the hyperfine transition frequency (F = 1, m_F = 0) → (F = 0, m_F = 0) of the ground state of the hydrogen atom [12, 15, 22]. According to the data of the latest measurements [24], it is equal to 1,420,405,751.800 + 0.028 cps; this value was obtained on the hypothesis that the frequency of the cesium standard $\Delta\nu$(Cs) = 9,192,631,770.0 cps. The precise determination of the hyperfine transition frequency is intimately related to the attainment of high relative and absolute frequency stability in hydrogen oscillators. The best values that have been obtained to date [25] are as follows: relative frequency stability of two oscillators $1.1 \cdot 10^{-13}$ after 15 sec, after 5 days about $5 \cdot 10^{-13}$, frequency reproducibility $2 \cdot 10^{-13}$ [26].

It is customary in tuning the frequency of the resonator to the resonance line to use modulation of the Q of the line. This modulation is realized by the addition of deuterium or oxygen atoms to the atomic beam [20, 24] or by increasing the intensity of the beam. In both instances the line is broadened, first due to the collisions of hydrogen atoms with deuterium or oxygen atoms, second due to exchange interaction between the hydrogen atoms. The criterion for tuning of the resonator to the frequency of the line is a lack of dependence of the oscillation line frequency on the beam intensity or the number of oxygen or deuterium atoms. For a precise determination of the hyperfine transition line, frequency corrections are introduced to account for the frequency shift of the line in the magnetic field, for the second-order Doppler shift, and for the shift due to interaction of the atoms with the wall.

The hydrogen oscillator is an excellent frequency standard, but it can also be used to perform a number of interesting physical experiments [20, 24]. It is possible, with the aid of the hydrogen oscillator, to measure the magnitude of the hyperfine splitting for deuterium and tritium with better than previously attained accuracy. An investigation has already been conducted on the effect of an external electric field on the hyperfine transition frequency of the hydrogen atom [25].

When the intensity of the atomic beam is increased, the linewidth increases due to spin-exchange interaction of the hydrogen atoms. This effect can be used to measure the cross section of the spin-exchange process. By adding other gases to the vessel, oxygen for example, some inference may be drawn on the basis of the reduction in lifetime of the hydrogen atoms in the vessel as to the mechanism of the interaction between these gases and atomic hydrogen. In a strong external magnetic field it is possible to measure the magnetic moment of the proton.

A hydrogen quantum oscillator having high stability can be used to perform experiments to test the theory of relativity.

The present article is devoted to the design problems, the choice of optimum parameters for the device, and an investigation of the operation of quantum oscillators utilizing a beam of hydrogen atoms.

CHAPTER 1

DESIGN AND ADJUSTMENT OF THE HYDROGEN ATOMIC BEAM QUANTUM OSCILLATOR

1. Operating Principle of the Hydrogen Atomic Beam Quantum Oscillator

A quantum oscillator utilizing a beam of hydrogen atoms (frequency ν_0 = 1420.405 Mc) operates on the hyperfine splitting transition (F = 1, m_F = 0) $\rightarrow$ (F = 0, m_F = 0) of the hydrogen atom. The hyperfine structure level diagram of the hydrogen atom is shown in Fig. 1.

The transition frequency depends slightly on the external magnetic field in the weak field region and is determined according to the formula $\nu = \nu_0 + 2750H^2$ (H in oersteds).

A schematic diagram of the apparatus is shown in Fig. 2. At the source of atomic hydrogen the hydrogen molecules are disintegrated by electrical discharge into atoms; the atoms pass through a small orifice into a region of high vacuum. The diaphragm is used to cut out a narrow atomic beam, which is transmitted along the axis of the focusing six-pole magnet. The atoms occurring in the states F = 1, m_F = 0 and F = 1, m_F = 1 are focused onto the entrance

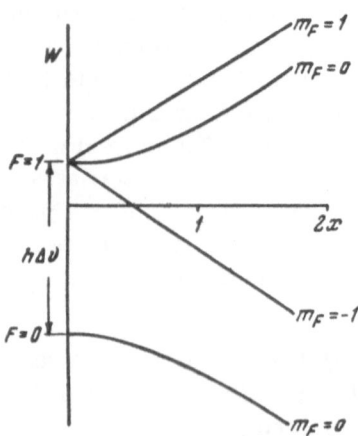

Fig. 1. Hyperfine structure level diagram for the ground state of the hydrogen atom in a magnetic field: $x = -(g_J - g_I)\frac{\mu_0 H_0}{h\Delta\nu}$.

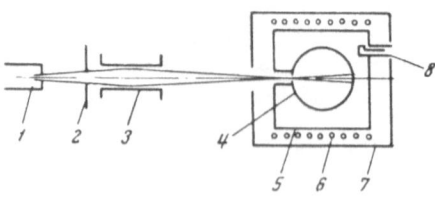

Fig. 2. Schematic diagram of the hydrogen quantum oscillator. 1) Source of atomic hydrogen; 2) diaphragm; 3) focusing magnet; 4) quartz vessel coated internally with teflon; 5) resonator cavity; 6) solenoid for generation of axial magnetic field; 7) magnetic shield; 8) coupling loop.

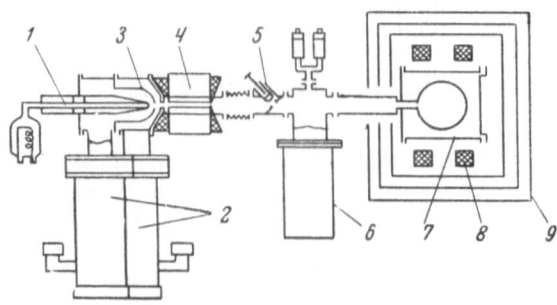

Fig. 3. Diagram of apparatus No. 1. 1) High-frequency source; 2) N-5SM-1 pump; 3) diaphragm; 4) six-pole electromagnet; 5) vacuum valve; 6) GIN-0.5 ion getter pump; 7) cavity with vessel; 8) Helmholtz coil; 9) magnetic shields.

orifice of the storage vessel, which is placed in the center of a cavity resonator tuned to the frequency of the transition (F = 1, m_F = 0) → (F = 0, m_F = 0). The inside walls of the storage vessel are coated with teflon (Ftoroplast-4); essentially the atoms do not change their quantum state in collision with the wall. The entrance orifice of the vessel is made small enough for an atom to be present in the vessel (and, hence, to interact with the high-frequency field), on the average, for a period of the order of a second. In connection with such a long interaction time the emission line of the atomic hydrogen turns out to be narrow (about 1-2 cps).

The region in which the vessel is situated is protected by magnetic shields against the earth's and other stray magnetic fields. A weak constant uniform magnetic field is imposed in the direction of the magnetic vector component of the high-frequency field.

2. The Vacuum System

Each of the apparatuses described above consists of three vacuum chambers, interconnected through an opening for the transmission of the atomic beam; the chambers are evacuated with high-vacuum pumps. A diagram and general view of the apparatus are shown in Figs. 3-5.

The apparatuses differ in the construction of the third section, namely the one nearest the cavity; the first two sections are identical. They comprise tubes 140 mm in diameter, welded from stainless steel sheet with flanges of the same steel. The tubes are connected to diffusion pumps by vacuum tubes with a square cross section. The source section is separated from the intermediate chamber by a copper bell-shaped partition, in which a hole 3 mm in diameter is drilled on the beam axis. This particular wall configuration is dictated by the need for making the source aperture as near as possible (in order to increase the capture angle of the focusing magnet) to the magnet entry, without deteriorating the gas evacuation conditions too appreciably.

The source holder is attached to the system through an alignment device. The source holders are made in two versions, for a Wood's

Fig. 4. General view of apparatus No. 1 (magnetic shields removed).

Fig. 5. General view of apparatus No. 2 (magnetic shields removed).

source and a high-frequency source. The flange of the six-pole sorting magnet is attached to
the intermediate chamber. The pole shoes are soldered into a brass tube 80 mm in diameter
so as to facilitate evacuation. The coils of the magnets are placed outside the vacuum system.
The magnet is connected to the third high-vacuum section through a sylphon bellows. The third
section of the first apparatus is T-shaped, with two flanges situated on the beam axis and, connected to it, a GIN-0.5 titanium ion getter pump [27]. A vacuum valve is located between the
second and third sections. With the valve closed and the pumps on, the pressure in the last section (including the vessel) does not go above $1 \cdot 10^{-4}$ mm Hg. The quartz vessel is evacuated
through a quartz tube 30 mm in diameter; the tube is tightened to the flange of the cavity with a

rubber ring. The cavity flange is connected in vacuum-tight fashion through a stainless steel tube to the third section. In the third section of apparatus No. 2 an N-5SM-1 vapor-oil pump is used. The ultimate vacuum in this section is $2 \cdot 10^{-7}$ mm Hg, which is better than the manufacturer's specifications for the pump. This improvement was made possible by means of a nitrogen trap, wherein the oil path along the "hot" wall is closed off by a teflon partition. There is no separatory vacuum valve between the second and third sections in apparatus No. 2.

Type N-5 oil pumps with a capacity of 500 liters/sec were used to evacuate the first two sections of the apparatus No. 2. The first tests showed, however, that these pumps are not very well suited to the evacuation of hydrogen. In order to obtain a satisfactory hydrogen pumping rate with these pumps, the power of the heater had to be approximately doubled (raised to 1.5-2 kW). Another drawback of these pumps is their rather narrow (by comparison with the TsVL-100 and N-5SM-1 pumps) discharge pressure, which could not always be provided by the RVN-20 forepumps that we used. It was necessary, therefore, to connect TsVL-100 pumps to the discharge tube of each N-5 pump as boosters. The N-5SM-1 pumps turned out to be more adaptable for the pumping of hydrogen. Of course, their pumping rate had to be increased by increasing (by 10-20%) the power of the heater. Normally in the operation of the quantum oscillator the hydrogen pressure at the source is equal to $2-4 \cdot 10^{-1}$ mm Hg. A flux $Q_1 = 1 \cdot 10^{18}$ atoms per sec is admitted through an orifice 0.8 mm in diameter into the first section of the vacuum system. The rate S_1 of hydrogen evacuation from the first section, as calculated with allowance for the resistance of the lines [28], amounts to about 200 liters/sec; the working hydrogen pressure in it is $p_1 = Q_1/S_1 = 1.5 \cdot 10^{-4}$ mm Hg. The gas pressure in the second and third sections of the working apparatuses is equal to $3 \cdot 10^{-6}$ and $3-5 \cdot 10^{-7}$ mm Hg, respectively.

The pressure in the sections where the vacuum was below $1 \cdot 10^{-7}$ mm Hg was measured with IM-11 [29] and LM-2 tubes. By measuring the vacuum with two tubes simultaneously, we calibrated the LM-2 tube in the region of pressures below $1 \cdot 10^{-7}$ mm Hg. It turned out that the readings of the LM-2 tubes systematically exceeded the readings of the IM-11 by $4 \cdot 10^{-8}$ mm Hg. The value of the current corresponding to a pressure of $4 \cdot 10^{-8}$ mm Hg is probably the value of the LM-2 photocurrent.

3. The Atomic Beam Source

Hydrogen atoms are admitted into the vacuum system through an aperture 0.8 to 1 mm in diameter in a thin glass wall. In this case, the atomic beam has a wide angular distribution. The ultimate pressure at which the eflux of atoms may be regarded as molecular is determined by the condition [30]

$$\lambda \geqslant d, \tag{1}$$

where λ is the mean free path of an atom in the source and d is the diameter of the aperture. For d = 0.8 mm this ultimate pressure is equal to 0.23 mm Hg. If the pressure in the source is raised above the critical value [as defined by the inequality (1)], the intensity of the beam continues to grow, but the rate of growth is slowed down in this case (the intensity is no longer proportional to the pressure) [30].

Multichannel collimators are often used for the creation of a directional atomic beam [21, 31-34]. The dissociation of the hydrogen molecules in the immediate vicinity of the source aperture takes place as a result of electron impact in sources with a glow discharge (Wood's tube and the high-frequency source). Dissociation can also be realized by heating the gas to a high temperature [35].

The steady-state concentration of hydrogen atoms [H] in the discharge zone may be determined from the equation

$$\left.\begin{array}{l} \dfrac{d\,[\mathrm{H}]}{dt} = k_1\,[\mathrm{H}_2]\,W - k_2\,[\mathrm{H}] - k_3\,[H]^2 = 0, \\[2mm] [\mathrm{H}_2] = [\mathrm{H}_2]_0 - \dfrac{1}{2}\,[\mathrm{H}]. \end{array}\right\} \qquad (2)$$

Here $[\mathrm{H}_2]$ is the concentration of molecular hydrogen, $[\mathrm{H}_2]_0$ is the same in the absence of discharge, W is the electrical power absorbed per cubic centimeter of gas in the discharge zone, k_1 is the yield of atomic hydrogen per unit absorbed power, k_2 is the recombination constant of atomic hydrogen at the walls of the source chamber, and k_3 is the recombination constant of atomic hydrogen in the volume.

The first term in Eq. (2) corresponds to the rate of formation of atomic hydrogen in the volume under the influence of discharge, the second and third terms correspond to the rate of annihilation of atomic hydrogen at the walls and in the volume, respectively. At gas pressures of about 0.1-0.5 mm Hg the annihilation of atomic hydrogen is determined primarily by recombination at the walls [36]. Consequently, the steady-state concentration of atomic hydrogen in the discharge zone, according to Eq. (2), is equal to

$$[\mathrm{H}] = \frac{k_1 W}{\dfrac{1}{2}\,k_1 W + k_2}\,[\mathrm{H}_2]_0. \qquad (3)$$

It is evident from the expression (3) that [H] can be increased by raising the discharge power W or by diminishing k_2.

The recombination constant of atoms at the walls can be decreased by means of special coatings. Substances that come highly recommended for coating of the inner surface of atomic hydrogen sources are the silanes [a mixture of dimethyl dichlorsilane $(CH_3)_2Cl_2Si$ and trimethyl chlorsilane $(CH_3)_3ClSi$], or the surface can be treated with metaphosphoric acid or a solution of potassium tetraborate $K_2B_4O_7$. The degree of dissociation of hydrogen in a source with treated walls can be raised to 95% [31, 32, 36].

For purification of the hydrogen and for regulating the degree of its influx into the vacuum system we used nickel flow accumulator (Fig. 6). The action of the nickel flow accumulator is based on hydrogen's 2000 times better capacity than other gases to diffuse through nickel heated to a high temperature (500°C) [37-39]. A thin-walled nickel tube 2.64 mm in diameter with a wall thickness of 0.1 mm is used in the accumulator. A tube about 150 cm long was first annealed, then the tube was filled with water, and the water was frozen by lowering the tube into liquid nitrogen. The ice-filled tube was wound on a mandrel 50 mm in diameter, and as the tube thawed it was again cooled with liquid nitrogen. The housing of the accumulator was made of molybdenum glass. Two Kovar tubes with an outside diameter of 10 mm were soldered into the glass 4 cm apart from one another. Terminals and small tubes for attachment of the nickel tube were bonded to the tubes with hard solder. For protection of the soldered region against high temperatures three turns of copper tubing were wound on each of the Kovar tubes, and water was circulated through the tubing. During operation the housing of the accumulator is fanned with a ventilator. The temperature of the housing in this case does not go above 100°C. In order to avoid the accumulation of contaminants inside the nickel tube, hydrogen was passed through it continuously. The power required for heating of the tube is about 150 W.

The hydrogen that diffuses through the nickel tube is admitted to the beam source. We tested sources using a glow discharge in a hydrogen atmosphere at a pressure of $1-5 \cdot 10^{-1}$ mm Hg. One of them was the well-known Wood's tube [40]. This comprises a molybdenum glass tube (Fig. 7), 10 mm in diameter, with a wall thickness of 1 mm and length of about 140 cm, divided in halves in the shape of a U-tube. The tube is contained in a glass sheath 36 mm in

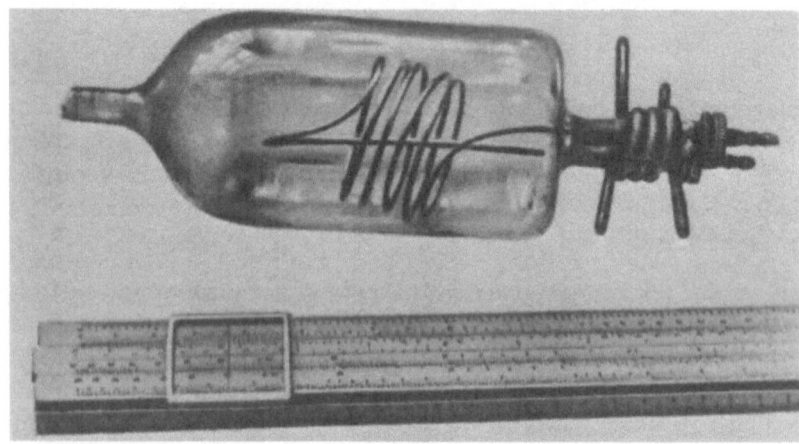

Fig. 6. Nickel flow accumulator.

Fig. 7. Wood's source.

diameter, through which water is circulated for cooling. In the middle of the tube, at the bend, a circular aperture is ground through 0.8 mm in diameter for formation of the atomic beam [41]. The ends of the tube are closed off with small bulbs 70 mm in diameter, into which are inserted water-cooled aluminum electrodes. A constant voltage of 4-6 kV is applied to the electrodes of the discharge tube, the current through the tube being equal to 200-300 mA. The inside surface of the Wood's tube was rinsed with a three-percent solution of potassium tetraborate prior to installation of the source in the system.

The discharge tube is secured in a special holder (see Fig. 7). The tube is sealed with two rubber rings. The ring nearest the source aperture takes care of the actual seal, while the ring on the opposite side of the tube keeps it from shifting under the influence of atmospheric pressure. Adjustment of the accessory comprising the two flanges between which the sylfon bellows is soldered permits the distance from the source aperture to the magnet entry to be varied and the source aperture to be displaced within certain limits, by changing the slope of the tube.

In addition to the Wood's source, we tested a high-frequency source, specifically a straight molybdenum glass tube 35 cm in length, with an outside diameter of 10 mm and wall thickness

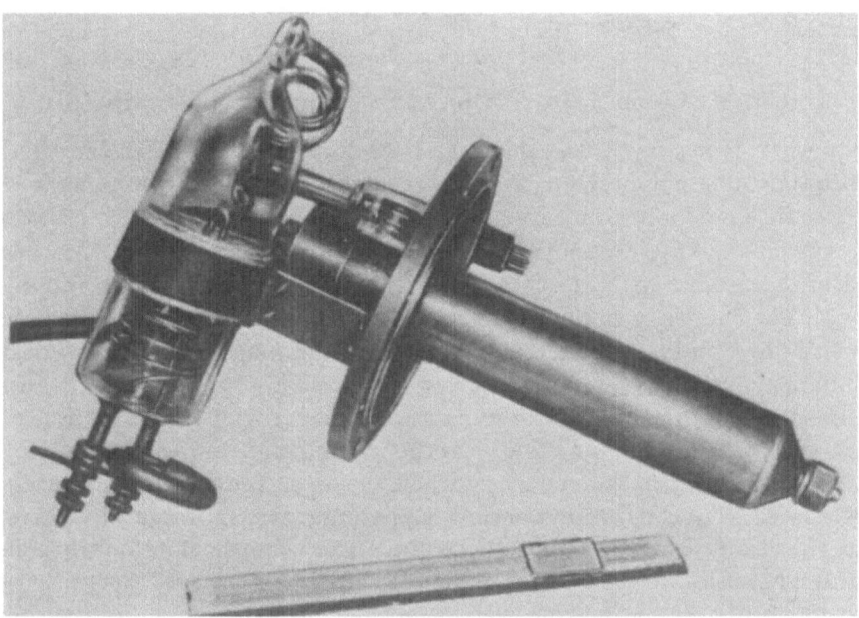

Fig. 8. High-frequency source.

of 1 mm. The tube was soldered closed at one end and ground off until a hole 0.8 mm in diameter was obtained; the other end of the tube was soldered to the nickel flow accumulator. A water-cooled annular electrode is seated on the tube, the casing of the tube holder functioning as the second electrode (Fig. 8). The tube is powered by a 30-Mc, 50- to 100-W high-frequency generator.

The intensity of the atomic beam was determined according to the rate of darkening (actually, the rate of turning blue) of molybdenum oxide chemical targets (see §5, Chapter I). We could not detect any appreciable difference in the intensities of the atomic beams created by the two types of sources.

4. Sorting by States and Focusing of the Atomic Beam

For the removal of hydrogen atoms in the states $F = 1$, $m_F = -1$, and $F = 0$, $m_F = 0$ from the beam and focusing of atoms in the states $F = 1$, $m_F = 0$ and $F = 1$, $m_F = +1$ onto the entrance aperture of the storage vessel, the quantum oscillator utilizes the tendency of atoms in different states to suffer different deviations in the inhomogeneous field of the six-pole magnet. The possibility of focusing a beam of molecules having a magnetic dipole moment by means of the field of a six-pole magnet was first indicated by Korsunskii and Fogel' [42], as well as by Friedburg and Paul [43, 44]. Subsequently, many authors [45-47] had occasion to discuss the operation of focusing magnets.

We wish to consider the behavior of an atom having a magnetic dipole moment μ and falling into an inhomogeneous magnetic field.

As apparent from Fig. 1, the additional energy acquired by a hydrogen atom in the states $F = 1$, $m_F = \pm 1$ in a magnetic field of strength H is described by the equation

$$W = -\mu \mathbf{H}. \tag{4}$$

The force acting on an atom in an inhomogeneous field is

$$\mathbf{F} = -\nabla W = -\frac{\partial W}{\partial H}\,\nabla H = \mu_{\text{eff}}\nabla H, \tag{5}$$

where μ_{eff} is the projection of $\boldsymbol{\mu}$ on the direction of the field $\mathbf{H}$, $\mu_{\text{eff}} = \neg dW/dH$.

For the states F = 1, $m_F = \pm 1$, the quantity μ_{eff} does not depend on the field, and its modulus is equal to the Bohr magneton μ_0. For the states F = 1, $m_F = 0$ and F = 0, $m_F = 0$, the effective magnetic moment depends on the external magnetic field. For H = 0 we have $\mu_{\text{eff}} = 0$. As the field is increased, the atom acquires a magnetic moment $\mu_{\text{eff}} \neq 0$. In strong fields the magnetic moment of the atom μ_{eff} is determined primarily by the magnetic moment of the electron and is near μ_0. The increase in μ_{eff} with increasing field occurs rather quickly, becoming practically equal to μ_0 in a field of about 1000 Oe. Since the magnetic field strength in the focusing magnets of the apparatus is relatively large (greater than 1000 Oe), we will assume for simplicity in the ensuing calculations that the magnetic moment of the atom $\mu_{\text{eff}} \simeq \mu_0$ for the states F = 0, $m_F = 0$, F = 1, $m_F = 0$ as well. On the other hand, in order for it to be permissible to consider μ_{eff} constant, it is necessary when the atom moves in a variable magnetic field to retain the same value for the projection of the magnetic moment of the atom on the direction of the external field vector. This will be the case when the Larmor precession frequency ν_L of the atomic magnetic moment in the magnetic field is much larger than the frequency $\overline{\Omega}_0$ of the variable external field. An atom whose motion takes it into a field with a different strength perceives that field as an alternating field, varying with a frequency

$$\overline{\Omega}_0 = \frac{\overline{v}}{l_2}, \tag{6}$$

where $\overline{v}$ is the mean velocity of the atoms and l_2 is the length of the magnet. Substituting the numerical values of the variables in (6), we find that $\overline{\Omega}_0 \sim 10^4$ cps.

The Larmor precession frequency [48] is equal to

$$\nu_L = \frac{[\boldsymbol{\mu}_J \mathbf{H}]}{2\pi\hbar\mathbf{I}} = \frac{g_J\mu_0 H}{h} = 1.4 g_J H\,(\text{Mc}) \tag{7}$$

where g_J is the atomic g-factor ($g_J = 2$ for the $^2S_{1/2}$ state), and H is the field strength in oersteds.

In a field H $\simeq$ 1000 Oe, the frequency $\nu_L = 2.8 \cdot 10^9$ cps, i.e., it is much greater than $\overline{\Omega}_0$. This implies the constant of μ_{eff}.

For focusing of atoms with a negative μ_{eff}, it is necessary to pick a magnetic field configuration such that

$$\mathbf{F} = -D\mathbf{r}, \tag{8}$$

where $\mathbf{r}$ is the distance measured from the beam axis and D is a constant.

A two-dimensional (plane) magnetic field can, according to potential theory, be represented as the gradient of the real part of an analytic function $f(x + iy)$, representing the complex potential [42-44].

The requirement (8) is satisfied by a magnetic field described by the complex potential

$$f(x + iy) = c(x + iy)^3. \tag{9}$$

The equations for the force and equipotential lines of the field have the form

$$U = c(x^3 - 3xy^2), \tag{10}$$

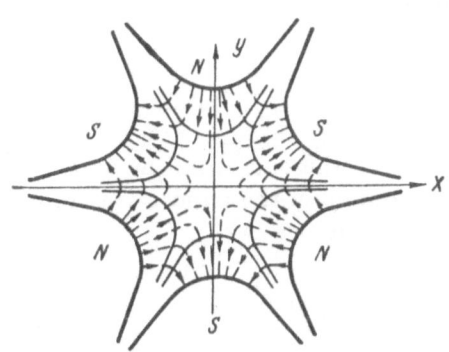

Fig. 9. Lines of force and equipotentials of the focusing magnetic field described by Eq. (14). The lines of force are indicated by dashed lines; the poles by heavy lines.

$$V = c\,(3x^2 y - y^3). \tag{11}$$

Differentiating $\operatorname{Re} f(x + iy)$ with respect to the coordinates, we find

$$\mathbf{H} = \operatorname{grad} \operatorname{Re} f\,(x + iy) = \operatorname{grad} \operatorname{Re}\,[c\,(x + iy)^3]; \tag{12}$$

$$H_y = -6cxy, \quad H_x = 3c\,(x^2 - y^2), \quad H_z = 0. \tag{13}$$

Denoting $a \equiv 6c$, we obtain

$$H_x = a/2\,(x^2 - y^2), \quad H_y = -axy, \quad H_z = 0. \tag{14}$$

The force lines and equipotentials of the field are shown in Fig. 9. In a cylindrical coordinate system

$$\mathbf{r} = \mathbf{i}x + \mathbf{j}y, \quad r^2 = x^2 + y^2, \tag{15}$$

$$|H| = (H_x^2 + H_y^2)^{1/2} = \frac{a}{2}\,(x^2 + y^2) = \frac{a}{2}r^2. \tag{16}$$

Consequently, the force acting on the atom in such a field is equal to

$$\mathbf{F} = \mu_{\text{eff}} \nabla H = \mu_{\text{eff}}\, a\mathbf{r}. \tag{17}$$

This expression coincides with (8) if we set $\mu_{\text{eff}} a \equiv -D$. The magnetic field illustrated in Fig. 9 can be created by means of six poles of alternating polarity equally spaced from the center. The surface of the poles is described by the equation for the equipotential surface for a definite value of the potential. The poles are indicated in Fig. 9 by a heavier line. Estimates show that the shape of the pole shoes is not of paramount importance [47]. It is simpler from the engineering point of view to fabricate the pole shoes with a trapezoidal cross section. Lemonick, Pipkin, and Hamilton [46] tested a magnet whose pole occupied an angle $\vartheta = 40°$ on a cylinder with a radius r_0; 20° was given over to the gap between adjacent poles. Christensen and Hamilton [47] showed that an angle $\vartheta = 30°$ is more nearly optimal, because the field strength at the pole turns out to be higher in this case than for $\vartheta = 40°$, due to the reduction in the scattered field in the gap between poles.

The field described by Eqs. (10) and (11) can also be generated by means of six cylindrical poles with a current [44].

Let us consider the motion of an atom with an effective magnetic moment μ_{eff} in the field described by Eqs. (14). Let the source aperture (which we will regard as a point opening) lie on the z axis at a distance l_1 from the magnet entry (the z axis coincides with the axis of symmetry of the magnet), and let l_2 represent the length of the magnet, L the distance between the end of the magnet and the point of intersection of the atom's trajectory with the z axis (position of the "image" of the source aperture). We assume the initial velocity v of the atom to be equal to the most probable velocity in the beam $v_B = 1.22\sqrt{2kT/m}$, the velocity vector forming an angle α with the z axis (Fig. 10).

The equation for the trajectory of the atom on emerging from the magnet [42] has the form

$$r = \alpha\,(l_1 + z)\cos\frac{\omega l_2}{v} + \left(\frac{v}{\omega} - \frac{\omega l_1 z}{v}\right)\alpha \sin\frac{\omega l_2}{v}, \tag{18}$$

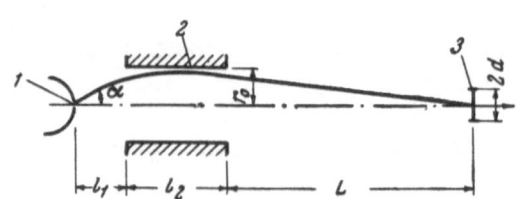

Fig. 10. Focusing diagram of the atomic beam. 1) Source; 2) focusing magnet; 3) entrance aperture of the storage vessel.

the trajectory intersecting the z axis where the radial coordinate goes to zero:

$$0 = \alpha (l_1 + z) \cos \frac{\omega l_2}{v} + \left(\frac{v}{\omega} - \frac{\omega l_1 z}{v} \right) \alpha \sin \frac{\omega l_2}{v}. \quad (19)$$

Hence the distance from the exit from the magnet to the point of intersection of the atom's trajectory with the z axis is equal to

$$L = \frac{l_1 \cos \frac{\omega l_2}{v} + \frac{v}{\omega} \sin \frac{\omega l_2}{v}}{\frac{l_1 \omega}{v} \sin \frac{\omega l_2}{v} - \cos \frac{\omega l_2}{v}}. \quad (20)$$

For apparatus No. 1, the distance from the source to the magnet $l_1 = 4$ cm, the length of the magnet $l_2 = 15$ cm, the distance between opposite poles $2r_0 = 0.32$ cm. The most probable velocity of the atoms in the beam at T = 300°K is $v_B = 1.22\sqrt{2kT/m} = 2.7 \cdot 10^5$ cm/sec, the cyclic frequency of radial vibrations of the atom in the magnet is

$$\omega = \sqrt{\frac{2\mu_0 H_0}{r_0^2 m}} = 6.85 \cdot 10^2 \sqrt{H_0} \text{ sec}^{-1}.$$

The magnet was tested with a distance of 32 cm from the exit of the magnet to the chemical target. The maximum beam intensity was recorded with a current of about 2 A in the windings of the electromagnet; according to Eq. (20), the field strength in this case, $H_0 \approx 1500$ Oe.

For apparatus No. 2, $l_1 = 7$ cm, $l_2 = 15$ cm, $r_0 = 0.3$ cm, the distance from the magnet to the target was equal to 80 cm. The maximum darkening of the target under the influence of the beam is observed with a current of 6-8 A in the windings of the electromagnet, corresponding to $H_0 \approx 3600$ Oe.

The capture angle α_{max} of the focusing magnet (i.e., the angular aperture α_{max} of the beam of atoms, ejected by the source, that are capable of being focused) is determined by the stipulation that an atom with a kinetic energy $mv^2/2$, emerging from the source at an angle α_{max} with respect to the axis of the magnet, can be returned to the axis by the magnetic field [30, 42, 46]. The atom is returned at the point where the following condition is fulfilled:

$$\frac{mv_\perp^2}{2} = \mu_0 H_0. \quad (21)$$

Here $v_\perp$ is the component of the velocity of the atom in the plane perpendicular to the z axis. Clearly, $v_\perp = v\alpha_{max}$. If we take for v the most likely velocity $v_B = \sqrt{3kT/m}$, the condition (21) may be written in the form

$$\alpha_{max}^2 = \frac{2\mu_0 H_0}{3kT} \quad (22)$$

or in terms of the solid angles

$$\Omega = \pi \alpha_{max}^2 = \frac{2\pi}{3} \frac{\mu_0 H_0}{kT}. \quad (23)$$

Thus, in order to increase the capture angle, it is necessary either to increase the magnetic field strength H_0 or to lower the temperature T of the gas in the source. Inasmuch as the entrance aperture of the vessel occupies only a portion of the cross-sectional area of the beam, it is likely that only a fraction of the atoms entrapped by the focusing magnet get into the vessel. This fraction is conveniently expressed in terms of the effective capture angle of the magnet,

Ω_{eff} [12]. It is clear that an entrance aperture with diameter 2d in the storage vessel will not only admit those atoms whose radial coordinates revert to zero at z = L, but also atoms (the velocity of which differs from the most probable beam velocity) whose radial coordinates fall within the interval [0, d] at z = L.

For convenience we write $\dfrac{\omega}{v} = \sqrt{\dfrac{2\mu_0 H_0}{3kT r_0^2}} \equiv \kappa$, whereupon Eq. (18) may be rewritten (for z = L) in the form

$$r(\kappa)|_{z=L} = \alpha\left[(l_1 + L)\cos\kappa l_2 + \left(\frac{1}{\kappa} - \kappa l_1 L\right)\sin\kappa l_2\right] \equiv \alpha A(\kappa). \tag{24}$$

An aperture of radius d will admit atoms that leave the source at an angle α with respect to the axis and have velocities in the interval $[v_{min}, v_{max}]$, where

$$\left.\begin{array}{l} r(v_{min})|_{z=L} = -d = \alpha A(\kappa_{max}), \\ r(v_{max})|_{z=L} = +d = \alpha A(\kappa_{min}). \end{array}\right\} \tag{25}$$

The plus and minus signs in front of d are arbitrary. The minus sign means that an atom having a velocity v_{min} and moving into the upper half-plane on emerging from the magnet intersects the z axis at z < L, while at z = L it is found in the lower half-plane at a distance d from the axis. The plus sign means that an atom having a velocity v_{max}, its trajectory lying in the upper half-plane and intersecting the axis at z > L, is found at a distance d from the axis (in the upper half-plane) for z = L. Then

$$r(v_{max}) - r(v_{min})|_{z=L} = +2d = \alpha[A(\kappa_{min}) - A(\kappa_{max})]. \tag{26}$$

This equation may be replaced by the expression

$$\frac{2d}{\alpha\Delta\kappa} = -\left.\frac{\partial A(\kappa)}{\partial\kappa}\right|_{\kappa=\kappa_B}, \tag{27}$$

where $\Delta K = K_{max} - K_{min}$, $K = K_B$ (K_B is the value of K for the most probable velocity in the beam),

$$\kappa_B = \frac{\kappa_{max} - \kappa_{min}}{2}.$$

Differentiating $r(K)_{z=L}$ with respect to K, we find

$$\frac{\partial}{\partial\kappa}r(\kappa)|_{z=L} = -\alpha\left\{\left[(l_1 + L)l_2 + \frac{1}{\kappa^2} + l_1 L\right]\sin\kappa l_2 + \left(\kappa l_1 L - \frac{1}{\kappa}\right)l_2\cos\kappa l_2\right\} \equiv \alpha M(\kappa). \tag{28}$$

The distribution of the atoms with respect to the velocities in the beam is described by the equation ([30], p. 26)

$$I(v) = \frac{2I_0}{\alpha^4}v^3 e^{-\frac{v^2}{\alpha^2}}, \tag{29}$$

where I_0 is the total intensity of the beam, $I_0 = \frac{1}{4}n\bar{v}A_S$, $\alpha = \sqrt{2kT/m}$,

$$\frac{I(v)\,dv}{I_0} = \frac{2v^3}{\alpha^4}e^{-\frac{v^2}{\alpha^2}}dv. \tag{30}$$

By definition K is inversely proportional to the velocity v, and hence

$$\frac{dv}{v} = -\frac{d\kappa}{\kappa}. \tag{31}$$

Therefore,

$$\frac{I(v)\,dv}{I_0} = -\frac{2v^4}{\alpha^4}e^{-\frac{v^2}{\alpha^2}}\frac{d\varkappa}{\varkappa}. \tag{32}$$

We replace v with the value of the most probable velocity in the beam $v_B = \sqrt{3\alpha/2}$, whereupon

$$\frac{I(v)\,dv}{I_0} = -\frac{d\varkappa}{\varkappa}. \tag{33}$$

The fraction of atoms that emerge from the source at an angle α with respect to the axis and are focused onto the entrance aperture of the vessel is determined by integrating I(v)dv from v_{min} to v_{max}. This integral may be written as follows in the variable K:

$$I_f = \int_{v_{min}}^{v_{max}} I(v)\,dv = -\frac{I_0}{\varkappa_B}\int_{\varkappa_B+\frac{\Delta\varkappa}{2}}^{\varkappa_B-\frac{\Delta\varkappa}{2}} d\varkappa = +\frac{I_0}{\varkappa_B}\Delta\varkappa, \tag{34}$$

i.e., $\Delta K/K_B$ denotes the fraction of focused atoms. To obtain the total flux of atoms admitted to the entrance aperture of the vessel, I_f must be integrated over the solid angle $d\Omega$ from zero to $\Omega = \frac{2\pi}{3}\frac{\mu_0 H_0}{kT}$ [see Eq. (23)]:

$$I_T = \frac{I_0}{\pi}\int_{r_{min}}^{r_{max}} \frac{2\pi r\,dr}{l_1^2}\frac{\Delta\varkappa}{\varkappa_B} = \frac{I_0}{\pi}\Omega_{eff}. \tag{35}$$

Here r is the radial coordinate of the atom at the entry of the magnet and is related to the angle α by the expression $r = l_1\tan\alpha \approx l_1\alpha$. Transforming to the variable α and replacing ΔK by its full expression [according to Eq. (27)], we obtain

$$I_T = \frac{2I_0}{\pi\varkappa_B}\int_0^{\alpha_{max}}\pi\alpha\,\Delta\varkappa\,d\alpha = \frac{2I_0}{\pi\varkappa_B}\int_0^{\alpha_{max}}\frac{2\pi d}{[-M(\varkappa_B)]}d\alpha = \frac{I_0}{\pi}\frac{4\pi d\alpha_{max}}{\varkappa_B[-M(\varkappa_B)]}. \tag{36}$$

Substituting α_{max} from (22) into (36), we find

$$\Omega_{eff} = -\frac{4\pi d r_0}{M(\varkappa_B)} = \frac{4\pi d r_0}{[(l_1+L)l_2+1/\varkappa^2+l_1 L]\sin\varkappa l_2+(\varkappa l_1 L-1/\varkappa)l_2\cos\varkappa l_2}. \tag{37}$$

According to the resulting expression, Ω_{eff} may be increased by increasing the diameter d of the entrance aperture of the vessel, letting the source aperture approach the entry of the magnet (diminishing l_1) and letting the vessel aperture approach the exit of the magnet (diminishing L) or increasing r_0. It is scarcely advisable to increase Ω_{eff} at the expense of an increase in r_0; in order to attain field gradients comparable with those obtained with magnets having a smaller gap, it would be necessary to increase the field strength H_0, which would demand an appreciable increase in the physical size of the magnet. Clearly, it is convenient to pick r_0 between the limits 1 and 3 mm. The dependence of Ω_{eff} on l_1 must be taken into account when designing the apparatus (the source aperture must be placed as near as possible to the entry of the magnet; the optimum value is $l_1 = 0$). There is no point in increasing d without otherwise altering the construction of the entrance aperture of the vessel, because then the emission line would be broadened due to the decrease in the incumbency time of the atom in the storage vessel; however, it is possible to keep the latter time the same and to increase the cross section of the entrance aperture, provided a single channel or group of channels is set up at the entrance to the vessel. This problem is discussed in §6 of Chapter I.

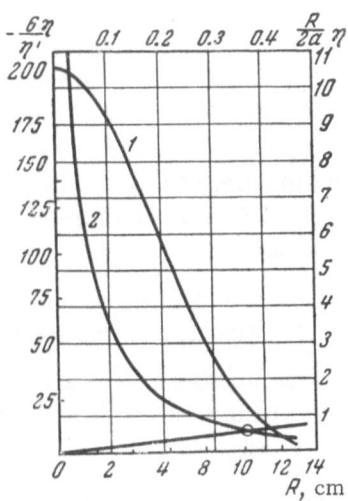

Fig. 11. The functions $\eta = <H_z>^2_V / <H^2>_r$ (1) and $-6\eta/\eta'$ (2) for a spherical storage vessel of radius R placed in the center of the cavity tuned to the mode TE_{011}, the diameter $2a$ of which is equal to the length h. The point of intersection of the graph of the function $-6\eta/\eta'$ with the graph of the function $y = R$ corresponds to the optimum (for self-sustained excitation) vessel radius R = 10.5 cm (R/$2a$ = 0.372).

Fig. 12. Six-pole electromagnet for focusing of the atomic beam.

Substituting the values $l_1 = 4$ cm, $l_2 = 15$ cm, L = 80 cm, $H_0 = 1500$ Oe, $r_0 = 0.16$ cm into Eq. (37), we find that for apparatus No. 1, $\Omega_{eff} \simeq 1 \cdot 10^{-4}$ ster. For apparatus No. 2, $l_1 = 7$ cm, $l_2 = 15$ cm, L = 116 cm, $H_0 = 3600$ Oe, $r_0 = 0.3$ cm, and $\Omega_{eff} \simeq 1 \cdot 10^{-4}$ ster.

The flux of atoms I_V in the state F = 1, $m_F = 0$ that are focused onto the entrance aperture of the vessel is equal to

$$I_V = \frac{1}{4} \frac{\Omega_{eff}}{4\pi} n\bar{v}A_s. \qquad (38)$$

With a pressure p = $1 \cdot 10^{-1}$ mm Hg at the source, assuming its temperature is equal to T = 300°C and the area of the exit aperture $A_s = 5 \cdot 10^{-3}$ cm^2, we arrive at

$$I_V = 6 \cdot 10^{12} \text{ atoms/sec.}$$

Let us estimate the flux of active atoms required for the oscillator to operate in a self-sustained mode [16].

The average power radiated by the atoms entering the vessel is

$$\Delta P = I\hbar\omega \frac{\theta^2}{1 + \theta^2 + \delta^2}, \qquad (39)$$

where

$$\theta^2 = \left(\frac{\mu_0}{\hbar}\right)^2 \frac{\langle H_z\rangle^2_V}{\gamma^2}, \quad \delta = \frac{\omega - \omega_0}{\gamma}.$$

It is helpful at this point to introduce the quantity η, representing the ratio of the average energy stored in the vessel to the average energy stored in the cavity, $\eta = <H_z>^2_V / <H^2>_r$ (the form of η is shown in Fig. 11). Expressed in terms of η, θ^2 may be written in the form

$$\theta^2 = \left(\frac{\mu_0}{\hbar}\right)^2 \frac{8\pi W}{\gamma^2 V} \eta.$$

The power consumption in the cavity is

$$\Delta P_{con} = \omega \frac{W}{Q}. \qquad (40)$$

From (39) and (40) we obtain the threshold active-particle flux required for self-excitation of the oscillator

$$I_{thr} = \frac{\hbar V \gamma^2}{4\pi\mu_0^2 Q\eta}. \qquad (41)$$

From $V = 1.5 \cdot 10^4$ cm^3, $\gamma = 3$ sec^{-1}, $Q = 2 \cdot 10^4$, $\eta = 3$ we obtain $I_{thr} = 1.7 \cdot 10^{12}$ atoms/sec.

The construction details of the magnet are clear from Fig. 12. The length of the poles is equal to 15 cm, the pole and armature of the magnet are made of Armco iron. The pole pieces are trapezoidal in shape. In the magnets (as in [47]), the width of the pole pieces is made equal to the width of the gap between adjacent ends of the poles. The gap between opposite poles is equal to 3.2 mm for the magnet of apparatus No. 1, 6 mm for the magnet of apparatus No. 2. The winding of each pole consists of 110 turns of PSO braided wire 1.95 mm in diameter. The windings are connected in series and are supplied from a 24-V battery.

The current is regulated with rheostats. The beam intensity (determined with chemical targets) at the vessel entrance aperture depends very little on field strength of the magnet.

Tests were also run with a six-pole permanent magnet with a length of 8 cm and separation of 3.2 mm between opposite poles. The degree of divergence of the focused beam was determined by means of targets placed at various distances from the magnet.

The magnet described here is not the optimum one. Lemonick, Pipkin, and Hamilton [46] have shown that the capture angle of the magnet can be increased by using, instead of a magnet with a gap $2r_0$ that is constant over its length, magnets in which the separation of the opposite poles $2r_0$ increases from the entry to the exit of the magnet. The use of magnets of this type in quantum oscillators permits a reduction in the total flux of atoms from the source into the vacuum chamber while preserving the same beam intensity at the entrance to the vessel.

5. Detection of the Beam of Hydrogen Atoms; Techniques for Adjustment of the Device

Prior to final assembly of the apparatus (installation of the storage vessel), we conducted some preliminary experiments to observe the focused beam of atomic hydrogen. For the hydrogen beam detector we usually used chemical targets of molybdenum oxide MoO_3 [30, 31].

When atomic hydrogen impinges on the target, a partial reduction of the oxide takes place:

$$MoO_3 + H \rightarrow Mo_4O_{10}(OH)_2.$$

When this happens, the color of the target changes at the incidence site of the beam; it goes from a yellowish white to blue. The remarkable attribute of the chemical target is its ability to react only to atomic hydrogen; molecular hydrogen can interact with the target for several hours, literally without leaving a trace, whereas the trace of atomic hydrogen shows up after 0.5 to 2 min. The rate of development of the image can be used to estimate the intensity of the atomic beam. It was noted that a flux of 10^{14} atoms/sec $\cdot$ cm^2 produced a developed image on the target after a one-minute exposure [31]. The intensity of the atomic beam at a distance of about one meter from the exit of the magnet is therefore equal to $5 \cdot 10^{13}$ to $2 \cdot 10^{14}$ atoms/sec $\cdot$ cm^2.

The targets were produced by burning 0.5-mm molybdenum wire in air. We usually used fresh targets, deposited on the same day the exposure was to be made. To obtain several traces of the beam on a single target, the molybdenum oxide was sprayed onto a glass strip, which was placed in a glass tube making up part of the vacuum system. The strip was cemented to a steel backing; the target could be manipulated with a magnet. The quality of the spots obtained on the plate was entirely acceptable, though not as good as on a single target situated on the inside of the vacuum chamber flange. This is attributable to the illumination of the target by scattered atoms, which tend to "live" for some time in the volume with glass walls (the probability of recombination of atoms at the wall is fairly small, about 10^{-3}).

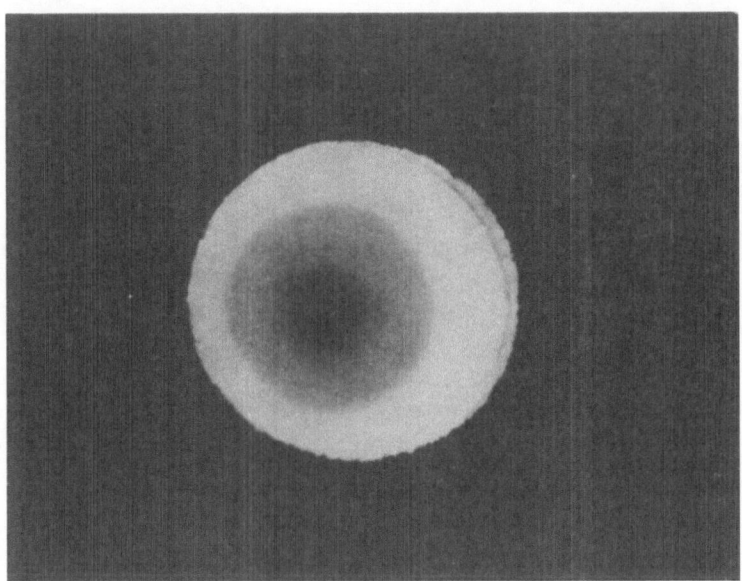

Fig. 13. Trace of the atomic beam on a chemical target.
The exposure time was 10 min.

A target having recorded a beam of atomic hydrogen is shown in Fig. 13.

The preliminary adjustment of the apparatus was accomplished as follows. An optical tube was mounted on the axis of the system, on the side where eventually the quartz vessel was to be installed. Our criterion of proper placement of the tube on the axis of the magnet was a uniform illumination of the poles of the magnet with an electric lamp placed on the same side as the beam source. After this, the adjustment accessory was used to align the source aperture symmetrically with respect to the poles of the magnet. This completed the operation of aligning the source.

The final adjustment of the system was accomplished by means of chemical targets. One version consisted in placing fine cross wires on the beam axis. Then the target was removed, and the center of the vessel aperture was aligned with the axis described by the source aperture and the intersection of the cross wires. The cross wires were removed once the adjustment was completed.

A second method proceeded as follows. The image of the beam was transferred from the target to a glass plate of the same size (the area of darkening was outlined with a circle). The plate was placed in the same position and in the same orientation as the target. It was possible to observe, with the aid of a telescope, whether the beam axis coincided with the axis of the magnet's poles. The discrepancy was eliminated by moving the source aperture in the appropriate direction. We usually used the second method as the simpler of the two.

The final adjustment of the system was performed on the basis of an induced emission signal or an oscillation signal. The increase in the signal amplitude served in this case as a criterion of the degree to which the adjustment was improved. It was noted that the amplitude of the induced emission signals varied only slightly with a rather appreciable displacement of the vessel aperture in the plane perpendicular to the beam. Consequently, the induced emission signal appears immediately after installation of the vessel, and a preliminary determination of the beam axis by means of a target is not required.

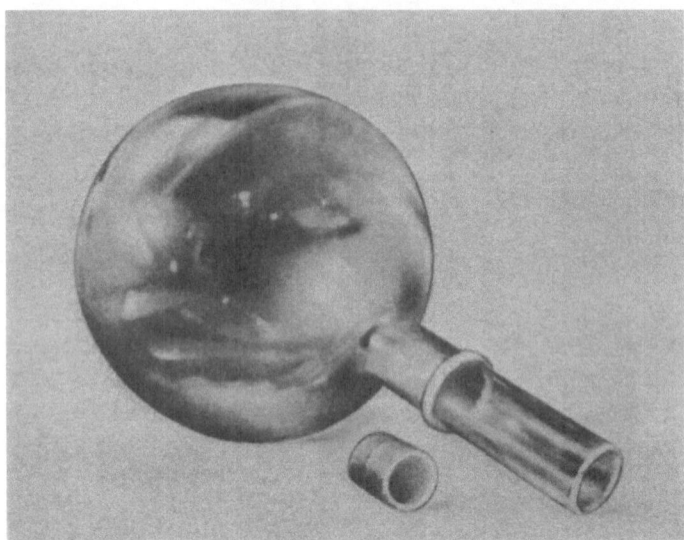

Fig. 14. General view of the storage vessel. The insertable diaphragm is lying alongside the vessel.

An independent criterion of proper adjustment of the source aperture is the variation in pressure of the hydrogen in the third section of the vacuum system. The total intensity of the atomic beam is $1/4$ of the diffusion flux of atoms into the third section (the 25% change in pressure with aligned and misaligned sources is easily detected). Normally, with maladjustment of the source the pressure in the third section was $4 \cdot 10^{-7}$ mm Hg, whereas with proper adjustment it dropped to $5 \cdot 10^{-7}$ mm Hg. This method of adjusting the source is used in the event a new source is installed without disassembly of the system and under conditions otherwise such that optical adjustment is impracticable.

6. The Vessel for Storing of the Atomic Hydrogen

The containing vessel performs two functions: It provides a base for the teflon film and is part of the vacuum system (hence its walls must be adequately thick, so as to withstand atmospheric pressure when the cavity is not under vacuum). A spherical vessel 17 cm in diameter with a wall thickness of about 1 mm, made of fused quartz, is used in the apparatus. Quartz was decided upon as a material for the vessel so that when placed in the cavity it would not deteriorate the Q of the latter too appreciably. The value of the function η is 3 for a vessel 17 cm in diameter (see Fig. 11). The vessel is held in the cavity on a quartz tube 30 mm in diameter and 10 cm in length (Fig. 14). Near the place where the tube is soldered to the vessel an interchangeable diaphragm is inserted in the tube on a ground-glass joint. Tests were conducted with a diaphragm 2 mm in diameter, drilled through a wall with a thickness of about 0.5 mm.

Teflon is the best of the currently known materials used to coat the storage vessels. The probability of recombination of hydrogen atoms on a teflon surface is nearly 10^{-5} [21, 23, 25]. An F-4D suspension was used to coat the inner surface of the vessel. The F-4D suspension is an aqueous suspension of disperse teflon particles with a grain size ranging from 0.05 to 0.5 μ [49, 50]. These particles are hydrophobic, as a result of which their aqueous dispersion is aggregate-unstable. To prevent the suspension from coagulating, surface-active agents are added to it. Prior to use, the suspension is diluted such that the polymer concentration is brought to

50%. A 60% suspension has a density of 1.5 g/cm^3, hence it is diluted with distilled water in the proportion of 30 cm^3 of water to 100 cm^3 of initial suspension. Before dilution the suspension is filtered several times through Kapron (similar to nylon) fabric.

Before being coated, the inner surface of the vessel is cleaned with hot bichromate, carefully rinsed with water, and dried by evacuation with a forepump. During evacuation the vessel is separated from the pump with a nitrogen trap, which serves to trap water vapors and to keep oil vapors from the pump from entering the vessel. After drying, about 50 cm^3 of the diluted suspension are poured into the vessel. The suspension has a great tendency to foam, which means that it must be poured very slowly, taking care to let it run along the wall. By slowly rotating the vessel, the suspension wets the entire inner surface. Then the vessel is set with the neck downward, so that the excess suspension will run out of the vessel. The time this takes is about 30 to 40 min. After waiting the required period of time, the vessel is dried by evacuation. It is important to remember that the evacuation rate must not run too high, otherwise the liquid might boil. The dried vessel acquires a white color. Normally the thickness of the film that is left on the surface of the vessel is about 10 μ. If pits are present on the glass surface, the liquid may form droplike beads on those places. If the thickness of the coating is greater than 60 μ (critical thickness) at such places, the film cracks. The cracks are easily detected when the film is examined under a magnifying glass. If cracks are detected in the film, the coating is washed away with distilled water (until it is baked, it is very infirm).

Once it has been assured that the coating is free of defects, baking of the film can proceed. The baking of the F-4D is begun at a temperature of 327°C, but it proceeds most rapidly at a temperature of 370 ± 10°C [49]. Care must be taken to keep the temperature within the prescribed limits, as teflon begins to disintegrate vigorously at temperatures above 400°C. During heating, the vessel is evacuated with an N-1S-M diffusion pump through a nitrogen trap. For this purpose the neck of the vessel is brought out through an opening in the cover of the furnace and is connected in vacuum-tight fashion to the vacuum system. During evacuation of the air from the vessel, the film is degassed, and the products of decomposition of the teflon, which are injurious to the health, can easily be conveyed outside the working premises.

The coating is normally baked for 30 min. After this, the temperature of the furnace is lowered to 340-360°C, and the vessel is outgassed for 3-4 h. Then the furnace is turned off and allowed to cool slowly with the vessel inside. Slow cooling promotes better adherence of the film to the wall. The insertable diaphragms are similarly coated; it is also possible to make the diaphragm of solid teflon [21].

The undegassed film has a yellowish hue, due to the presence of traces of surface-active agents in the film. Upon degassing, the film becomes whitish due to crystallization of the teflon. The quality of the coating was tested by introducing water droplets into the vessel. The absence of areas wetted by the water served as a good criterion of the quality of the surface. If defects were present in the film, the water flowed in under the film and gradually peeled it from the glass.

The vessel, coated internally with a film of teflon, is placed in the center of the high-frequency resonator cavity.

The average incumbency time $\bar{\tau}$ of atoms in a spherical vessel of radius R admitting a beam through an aperture in a thin wall is equal to

$$\bar{\tau} \equiv \gamma^{-1} = \frac{16\pi R^3}{3A_e\bar{v}} = \frac{4V_V}{A_e\bar{v}},$$ (42)

where V_V is the volume of the vessel, A_e is the area of the aperture, and $\bar{v}$ is the mean velocity of the atoms in the vessel.

The quantities γ and η [see (39)] depend on the radius of the vessel. It is to be expected, therefore, that there exists a certain value of the vessel radius for which I_{thr} (41) is a minimum.

We assume that $\gamma(R) \sim R^{-3}$. We substitute this expression into Eq. (41) and find the extremum. The extremal values of I_{thr} are obtained for values of R determined from the equation

$$R = -\frac{6\eta}{\eta'}. \tag{43}$$

The form of this function is shown in Fig. 11. Its intersection with the line $y = R$ yields the optimum radius of the vessel.

Due to the fact that the cross section of the beam at the place where it enters the vessel is circular with a diameter of 3-6 mm, it is possible to enhance the flux of atoms admitted to the vessel by increasing the entrance opening of the latter. In order not to diminish the lifetime of the atoms in the vessel in so doing, it is necessary to insert a duct of definite length in the entrance to the vessel. The average incumbency time of the atoms in the vessel with duct is equal to

$$\bar{\tau} = \frac{2R^3 l_d}{\bar{v}r^3}, \tag{44}$$

where l_d is the length of the duct and r is its radius. For example, for a vessel with a radius $R = 8$ cm and duct with $l_d = 8$ cm and $r = 2.5$ mm, we have $\bar{\tau} \approx 2$ sec. If a bundle of parallel ducts is inserted at the entrance to the vessel, we have

$$\bar{\tau}_{sys} = \frac{2\pi R^3 l_d}{\beta n A_e r' \bar{v}}, \tag{45}$$

where n is the number of ducts, r' is the radius of one duct, A_e is the total end area of the sheaf, and β is the transmission coefficient of the end. For a system of seven tightly packed tubes having an outside diameter of 2.6 mm and an inside diameter of 1.8 mm, contained in a tube 7.7 mm in diameter, the transmission coefficient β is about 0.5. Letting τ_{sys} be equal to 1 sec and $R = 8$ cm, we find $l_d = 1.5$ cm.

The flux of atoms into the vessel using a duct 5 mm in diameter is $(5/2)^2 \approx 6$ times the flux using a diaphragm with an aperture 2 mm in diameter. A group of ducts, even considering the 50% loss of the beam, increases the flux of atoms into the vessel by a factor of 5-10 over the flux transmitted by an aperture 2 mm in diameter.

The use of ducts at the entrance to the vessel is suitable for operation with low-intensity beams in the case when it is necessary to increase the containment time.

The end of the duct may be brought outside the cavity; by changing, for example, its length or the area of the entrance aperture, Q-modulation of the line can be effected (this is required in tuning the oscillator frequency to the frequency of the atomic transition line).

7. The Cavity Resonator

The cavity resonator of the hydrogen atomic beam quantum oscillator must have a sufficiently high Q and a stable natural frequency; the magnetic component of the high-frequency field must have an antinode at the place where the storage vessel is located.

The above requirements are best met by a cylindrical cavity resonator tuned to the wave mode TE_{011}. The magnitude of the longitudinal component of the magnetic field for the mode TE_{011} is a maximum on the cavity axis; on moving away from the axis, it decreases according to a $J_0'(\delta r)$ law, going to zero at the surface of a cylinder of radius $0.63a$ (where a is the radius of the cavity).

An important property of the TE_{011} mode is the absence of axial components of the current on the lateral surface of the cavity and of radial components on its base surfaces. For this reason, the surface discontinuities run along the current lines (for example, between the surface of the cylinder and the base) and do not result in deterioration of the Q. Consequently, the cavity may be easily tuned by providing the cylinder with a movable cover (leaving a space of 0.5-1 mm between the cover and the cylinder so as not to damage the surface of the cylinder when the cover is moved). The resonance wavelength of the cavity when tuned to the mode TE_{011} [51, 52] is equal to

$$\lambda = \frac{2a}{\sqrt{\left(\frac{\rho'_{01}}{\pi}\right)^2 + \left(\frac{\beta}{2}\right)^2}} = \frac{2a}{\sqrt{\left(\frac{3.832}{\pi}\right)^2 + \left(\frac{\beta}{2}\right)^2}}, \tag{46}$$

where $\beta = 2a/h$, h is the height of the cavity, ρ'_{01} is the root of the equation $J'_0(x) = 0$, $\rho'_{01} = 3.832$. For $\beta = 1$, the value of the no-load Q of the cavity in the mode TE_{011} passes through a maximum, where

$$\lambda = \frac{2a}{\sqrt{\left(\frac{3.832}{\pi}\right)^2 + \frac{1}{4}}}. \tag{47}$$

For a wavelength $\lambda = 21.1$ cm ($f = 1420.4$ Mc) we obtain from Eq. (47), $2a = h = 278.4$ mm.

For a cavity with silvered walls the theoretically attainable maximum Q can turn out to be about 80,000. The loaded Q of the cavity drops to 20,000-40,000 due to the presence of the apertures, the tuning rod, and the coupling loop.

A brass resonator with a diameter of 278.4 mm and mean length h = 280 mm was fabricated for operation. The length of the cavity could be varied by ±10 mm by shifting the position of the cover. The frequency of the resonator in this case varied within 20-Mc limits (the cavity was approximately retuned by 1 Mc by moving the cover 1 mm). The cavity was silvered and polished on the inside. For h = 2a = 278.4 mm the adjacent modes to TE_{011} are as follows (analytical data): TM_{012} (frequency of 1357.2 Mc), TM_{111} (1420.4 Mc) — the latter mode is degenerate with TE_{011}, and TE_{212} (1502.9 Mc). With the insertion of a spherical quartz vessel with a diameter of about 17 cm and wall thickness of about 1 mm into the cavity, the length of the cavity must be reduced to 247 mm in order to tune it to 1420.4 Mc (for the TE_{011} mode). The degeneracy of the modes TE_{011} and TM_{111} is removed with the vessel present in the cavity.

The frequencies of the modes TM_{111} and TE_{011} differ by about 6 Mc, thus enabling a high Q to be obtained for the mode TE_{011} due to the absence of intermodal coupling with TM_{111} [53]. The exact tuning of the cavity is realized by displacement of a rod 8 mm in diameter inserted into an opening located in the center of the cavity cover plate. At this location, the values of the magnetic and electric fields are nearly equal to zero, so that it is reasonable to expect very smooth tuning of the resonator frequency.

The influence of the depth of immersion of the rod on the resonance frequency of the cavity may be determined from the following relation ([51], p. 150; [54], II 26):

$$\frac{f - f_0}{f} = \frac{\Delta f}{f} = \frac{\mu \int\limits_{\Delta V} H'^2 dV - \varepsilon \int\limits_{\Delta V} E'^2 dV}{2\mu \int\limits_{V} H'^2 dV}. \tag{48}$$

The integration in the numerator is carried out over the volume ΔV occupied by the rod, in the denominator over the volume of the cavity V. Substituting the value of the fields E and H for the mode TE_{011} into (48) and carrying out the integration, we obtain

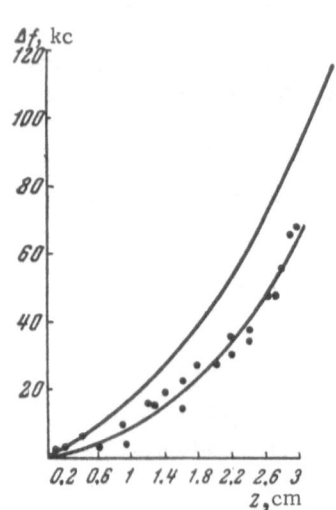

Fig. 15. Dependence of the natural frequency of the cavity on the depth of immersion z of the tuning rod.

$$\frac{\Delta f}{f} = \frac{\pi M(\delta r) z - \frac{h}{2} N(\delta r) \sin\left(\frac{2\pi}{h} z\right)}{\mu \pi a^2 h \delta^4 J_0^2(\delta a)\left(1 + \frac{\pi^2}{h^2 \delta^2}\right)}, \tag{49}$$

where

$$M(\delta r) \equiv \left(\varepsilon \omega^2 \mu^2 \delta^2 + \frac{\mu \pi^2 \delta^2}{h^2}\right) K(\delta r) + \mu \delta^4 L(\delta r),$$

$$N(\delta r) \equiv \left(\varepsilon \omega^2 \mu^2 \delta^2 - \frac{\mu \pi^2 \delta^2}{h^2}\right) K(\delta r) + \mu \delta^4 L(\delta r),$$

$$K(\delta r) = \frac{r^2}{2}\left\{\left[J_0(\delta r) - \frac{1}{r} J_1(\delta r)\right]^2 + \left(1 - \frac{1}{\delta^2 r^2}\right) J_1^2(\delta r)\right\},$$

$$L(\delta r) = \frac{r^2}{2}[J_1^2(\delta r) + J_0^2(\delta r)],$$

$\delta \equiv \rho'_{01}/a$, r is the radius of the rod, z is its depth of immersion, ε is the dielectric constant of the medium filling the cavity, and $\varepsilon \mu = 1/c^2$ (c being the velocity of light).

For a rod 8 mm in diameter, the tuning of a resonator operating in the mode TE_{011} with linear dimensions $2a = h = 278.4$ mm, according to (49), is described by the function

$$\Delta f = 1420.4 \cdot 10^2 [1.29 z - 5.16 \sin (0.226 z)] \text{ cps.} \tag{50}$$

The form of this function is shown in Fig. 15 (upper curve). The dots in the figure indicate the experimentally measured values of the retuning of the resonator as a function of the depth of immersion of the rod. All the points fall below the analytical curve; this is clearly attributable to the fact that the vessel distorts the form of the high-frequency field near the cover. For a resonator with a quartz vessel the strength of the high-frequency field near the cover and its rate of increase with z turn out to be lower than for the resonator without the vessel.

We need now to ascertain the influence of various parameters (dimensions of the cavity, pressure and moisture content of the air inside it) on the resonator frequency.

The relative change in wavelength to which the resonator is tuned is found by differentiation of the expression (46):

$$\frac{d\lambda}{\lambda} = \frac{4(\rho'_{01})^2}{4(\rho'_{01})^2 + (l\pi\beta)^2} \frac{da}{a} + \frac{(l\pi\beta)^2}{4(\rho'_{01})^2 + (l\pi\beta)^2} \frac{dh}{h}. \tag{51}$$

Here $\beta = 2a/h$. If $2a = h$, then $\beta = 1$. In this case, the expression is transformed to

$$\frac{d\lambda}{\lambda} = 0.855 \cdot \frac{da}{a} + 0.144 \cdot \frac{dh}{h}. \tag{52}$$

For a resonator made of a single material with a thermal expansion coefficient α in $\deg^{-1}$, we have

$$\frac{d\lambda}{\lambda} = -\frac{d\nu_r}{\nu_r} = \alpha dT°. \tag{53}$$

As shown in [16], the pulling of the oscillation frequency by the resonator may be described by the equation

$$d\nu_g = d\nu_r \frac{Q_r}{Q_l}. \tag{54}$$

Assuming $Q_l = 1.5 \cdot 10^9$ and $Q_r = 3 \cdot 10^4$, we obtain $d\nu_g = d\nu_r$, $2 \cdot 10^{-5}$.

If we are to attain an oscillation frequency stability no worse than $1 \cdot 10^{-13}$, $d\nu_r/\nu_r$ must be no greater than $5 \cdot 10^{-9}$. If the resonator is made of a material with an expansion coefficient $\alpha = 1 \cdot 10^{-7}$ deg^{-1}, it becomes necessary as a result to maintain the temperature of the resonator within limits $T = \pm 0.025°C$.

The dependence of the resonator frequency on the temperature can be abated considerably by constructing the cavity with thermal compensation [51]. The thickness d_0 of the compensating plate, which is made of a material with a coefficient of thermal expansion α_0 and is inserted into a cavity of length l made of a material with α_r, from Eq. (52) we find, letting $\alpha\lambda/\lambda = 0$,

$$d_0 = \frac{\alpha_r h}{0.144 \, (\alpha_0 - \alpha_r)} \, . \tag{55}$$

For a quartz cavity with $\alpha_r = 5 \cdot 10^{-7}$ (with $h = 250$ mm) the thickness of an aluminum compensating plate is $d_0 = 34$ mm.

The relative change in frequency of an air-filled cavity is equal (in absolute value) to the change in refractive index of air n [51]:

$$\frac{\Delta \nu_r}{\nu_r} = -\frac{\Delta n}{n} = -\Delta n. \tag{56}$$

The refractive index of air is given by the empirical formula

$$n = 1 + 1.05 \cdot 10^{-4} \frac{p}{T} + 50.3 \cdot 10^{-4} \frac{g\mathscr{E}}{T^2}, \tag{57}$$

where p is the pressure of the dry air in mm Hg, g is the relative moisture content of the air, $\mathscr{E}$ is the saturation vapor pressure at the given temperature, and T is the absolute temperature.

Differentiating (57) with respect to T, we obtain

$$\frac{\Delta n}{\Delta T} = \frac{1.05 \cdot 10^{-4}}{T}\left(\frac{\Delta p}{\Delta T} - \frac{p}{T}\right) + \frac{50.3 \cdot 10^{-4} g}{T^2}\left(\frac{\Delta \mathscr{E}}{\Delta T} - 2\frac{\mathscr{E}}{T}\right). \tag{58}$$

If we place the cavity in a hermetic sheath, $\Delta p/p = \Delta T/T$, so that the first term in the expression for $\Delta n/\Delta T$ goes to zero. Letting $T = 300°K$, $\Delta T = 5 \cdot 10^{-3}$ deg, the relative moisture content of the air $g = 50\%$, and, according to [51], p. 154, $\mathscr{E} = 23.55$ mm Hg and $\Delta\mathscr{E}/\Delta T = 16.77$ mm Hg, we obtain

$$\frac{d\nu_r}{\nu_r} = -\Delta n = -\frac{50.3 \cdot 10^{-4} g}{T^2}\left(\frac{\Delta \mathscr{E}}{\Delta T} - 2\frac{\mathscr{E}}{T}\right)\Delta T = 2 \cdot 10^{-8}, \tag{59}$$

which leads to a relative change in the oscillation frequency $d\nu_g/\nu_g = 0.7 \cdot 10^{-12}$. Consequently, in order to secure an oscillation frequency stability better than $1 \cdot 10^{-12}$, the cavity must be filled with a vacuum.

In tuning the cavity to the atomic transition frequency, care must be taken that the cavity is indeed tuned to the mode TE_{011}. As stated above, the neighboring modes are TM_{012} and TM_{111}, i.e., transverse magnetic-type modes. This means that one need only observe the particular type of mode in question. This is done by observing the power of a signal passing through the cavity on rotation of the coupling loop. In a number of instances the loop was replaced with a rod; then the signal vanished for the TE_{011} mode and remained present for transverse magnetic modes. Another criterion of correct tuning to the mode TE_{011} is the smoothness with which the cavity is retuned with the tuning stick (which is related to the weakness of the magnetic and

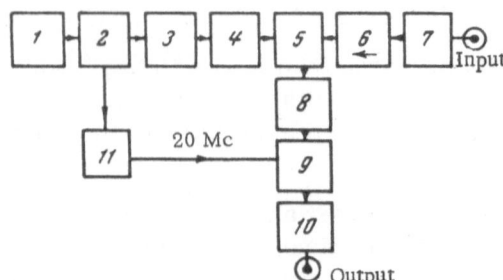

Fig. 16. Block diagram of a receiver operating at λ = 21 cm. 1) 1-Mc quartz oscillator; 2) hundredfold multiplier; 3) 1400-Mc
klystron converter; 4) klystron oscillator;
5) mixer; 6) ferrite decoupler; 7) noise
generator; 8) first ifa (f = 20.405 Mc, Δf =
1 Mc); 9) mixer; 10) second ifa (f = 405 kc,
Δf = 2.5 kc); 11) auxiliary multiplier

electric fields at the position of the rod). For
the modes TM_{012} and TM_{111} this dependence is
stronger (the resonator frequency is shifted by
about 30 kc when the rod is moved 1 mm, whereas, for the TE_{011} mode this shift requires the
rod to be moved 10 mm). The loaded Q of the
cavity with displacement of a coupling loop with
dimensions 10×14 mm^2 dropped from 32,000
(when the loop did not go beyond the plane of the
inner part of the cover) to 24,000 (when the loop
was completely inserted into the cavity). When
the center of the loop was removed to a distance
of 30 mm from the cover plane (the loop being
"recessed"), the value of the loaded Q was
38,000. It is apparent that the quality $Q_0 \simeq$
40,000 is determined by the energy losses in the
walls, the radiative losses through the second
coupling loop and aperture for admission of the
beam, and the losses in the walls of the quartz
vessel.

The value of the coupling quality Q_C is found from the relation

$$\frac{1}{Q_L} = \frac{1}{Q_c} + \frac{1}{Q_0},\qquad(60)$$

whence

$$Q_c = \frac{Q_L Q_c}{Q_0 - Q_L}.\qquad(61)$$

Assuming Q_L = 24,000, Q_0 = 40,000, we obtain Q_C = 64,000. The cavity of apparatus No. 2 had
dimensions $2a$ = 270 mm, h = 296.8 mm (with a vessel diameter of 168 mm), its load Q was
Q_L = 29,000 with the coupling loop in working position.

8. The Receiver for Emitted 1420-Mc Radiation

The power radiated by the atoms in the vessel is approximately equal to 10^{-13} to 10^{-12} W,
the coupling loop tapping a power of 10^{-15} to 10^{-14} W. For the detection of the generated signal
the sensitivity of the 1420-Mc radiation receiver must be at least $1 \cdot 10^{-15}$ W. Inasmuch as the
observation of induced radiation during the process of adjusting the oscillator requires the observation of even lower-power signals, it is desirable to have a receiver with a sensitivity of
about $5 \cdot 10^{-16}$ W. The necessary sensitivity is fairly easy to obtain, provided the receiver is
made with a sufficiently narrow pass band Δf [55]. We note that the emission frequency of the
hydrogen quantum oscillator is essentially fixed, hence it is possible in principle to design the
receiver with a rather narrow band.

For the power reception of a superhigh-frequency signal at a wavelength λ = 21 cm, we
used a superheterodyne receiver with double frequency conversion. The sensitivity of the receiver at 1420 Mc was approximately equal to 5-7 $\cdot 10^{-16}$ W. The first intermediate frequency
was equal to 20.405 Mc, the pass bandwidth of the first intermediate-frequency amplifier (ifa)
was 1 Mc. The second ifa had a frequency of 405 kc and pass bandwidth of 2.5 kc. The signal
power at the receiver input was measured by comparison with the known output of a noise generator. A block diagram of the receiver is shown in Fig. 16.

CHAPTER II

INVESTIGATION OF THE OPERATION OF THE HYDROGEN
ATOMIC BEAM QUANTUM OSCILLATOR (PRELIMINARY RESULTS)

1. Investigation of the Induced Emission of Atomic Hydrogen at 1420.4 Mc

The induced emission signal varies with time under the influence of an illumination pulse according to the following law, after the illumination is turned off [16]:

$$P_s = \frac{\omega \mu_0^2 \eta I^2 Q_L^2}{2\pi V \gamma^2 Q_c} e^{-2\gamma t},$$ (62)

where ω is the cyclic emission frequency of the atomic hydrogen, μ_0 is the Bohr magneton, I is the intensity of the atomic beam admitted to the storage vessel, Q_L is the load Q of the cavity, η is a function of a form shown in Fig. 11 (for a vessel 17 cm in diameter, $\eta = 3$), Q_c is the coupling Q, and V is the volume of the cavity.

By measuring the characteristic damping time of the radiation from the quantum oscillator, it is possible to determine the relaxation rate of the atoms and their lifetime in the excited state.

There are several processes contributing to relaxation of the atoms (broadening of the line) in the storage vessel. The total relaxation rate γ is comprised of the sum of the relaxation rates γ_i of these processes individually. The largest contribution comes from relaxation associated with the escape of active particles from the storage vessel (42.44) ($\gamma \sim 1$ sec^{-1}). The relaxation rate due to interaction with the walls is smaller (on the order of $7 \cdot 10^{-3}$ sec^{-1} for adiabatic collisions [16] and on the order of 0.7 sec^{-1} for nonadiabatic collisions [16]). Inhomogeneities of the magnetic field also cause relaxation of the atoms, primarily due to transitions between the Zeeman sublevels when the atoms move through an inhomogeneous field (about 0.5 sec^{-1} for a field inhomogeneity of 0.1 mOe [16]) and secondarily due to the disruption of coherence when the atoms move through an inhomogeneous magnetic field (this effect is small, and the relaxation rate is approximately 10^{-6} sec^{-1}).

Relaxation also occurs in connection with the collision of active hydrogen atoms with spin exchange. This rate is proportional to the number of hydrogen atoms in the vessel and is equal to 0.6 sec^{-1} for a hydrogen pressure of $\sim 5 \cdot 10^{-8}$ mm Hg. Relaxation and a concomitant line broadening also occur as a result of the first- and second-order Doppler effects, as well as due to collisions with residual gas atoms. The latter effect is the most pronounced but is minimized with good evacuation.

A block diagram of the apparatus for observing the induced emission of atoms is shown in Fig. 17. An 11.8-Mc signal is sent from the quartz oscillator to a multiplier. The oscillator frequency is determined by an evacuated quartz cavity placed in a thermostat. Variable capacitors are used to tune the oscillator. A 2^3-fold frequency multiplication occurs in three stages using 6Zh1P tubes. Then the resultant 94.4-Mc signal is transmitted through one 6Zh1P amplifier stage and arrives at a DKV-4 crystal diode. The fifteenth harmonic of the 94.4-Mc signal (1420.406 Mc) falls within the pass band of the cavity resonator ($\Delta f_r \simeq 30$ kc), which is tuned to the hyperfine transition frequency. The frequency multiplier is blocked by a negative voltage of -70V supplied to the tube grid and is triggered for a period of about 10 msec by a positive pulse from a multivibrator. The pulse repetition rate can be varied (by varying the RC) from 0.1 to 1.7 sec. Simultaneously, a pulse of negative polarization and the same duration partially

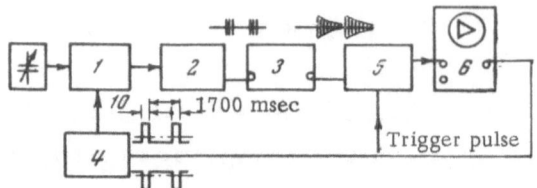

Fig. 17. Block diagram of the arrangement for observation of the induced emission of hydrogen atoms.
1) Quartz oscillator (f = 11.8 Mc); 2) 120-fold multiplier; 3) cavity of quantum oscillator; 4) modulator;
5) receiver; 6) ENO-1 oscilloscope.

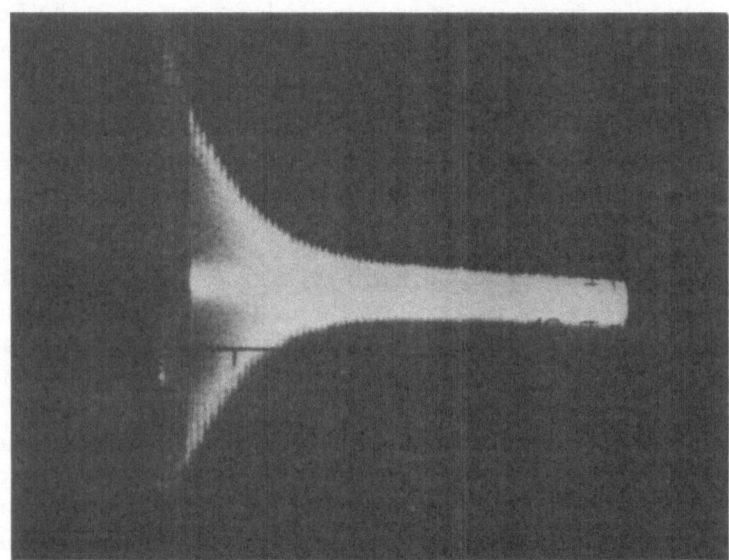

Fig. 18. Induced emission signal of atomic hydrogen after
stimulation by a light pulse. The pulse repetition period is
1.7 sec. The lifetime of the atoms in the vessel γ^{-1} = 0.3
sec.

blocks the receiver, serving to trigger the oscilloscope sweep. If atoms in the state F = 1,
m_F = 0 are present in the cavity, after termination of the pulse the power P_S is delivered to the
receiver [see Eq. (62)].

The signal voltage at the input is

$$u\,(t) = u_0 e^{-\gamma t}.$$

Assuming that there are no nonlinear distortions of the signal (this is true at low power levels),
we find that the shape of the signal envelope on the oscilloscope screen is described by the function

$$A\,(t) = A_0 e^{-\gamma t}.$$

Measuring the time for the signal amplitude to decay by a factor of $1/e$ on the oscillogram, one is able to determine the lifetime γ^{-1} of the atoms in the storage vessel. The error in determining γ^{-1} is 10%, γ^{-1} ranging from 0.03 to 0.3 sec, depending on the quality of the coating for different vessels.

Before proceeding with the experiment to determine the lifetime of the atoms in the vessel, we checked the resonator frequency. The cavity was tuned according to a quartz illumination oscillator, which in turn could be checked against a high-stability quartz oscillator.

A photograph of an oscillogram of the emission damping curve is shown in Fig. 18. It was obtained in the investigation of a teflon-coated vessel 17 cm in diameter, with an inside aperture diameter of 2 mm. The measured relaxation time was 0.3 sec and did not vary appreciably when vacuum in the vessel varied from $1 \cdot 10^{-7}$ to $1 \cdot 10^{-6}$ mm Hg.

By measuring the initial power $P_{s,in}$ of the induced emission signal, we were able to estimate the number of atoms in the upper state that were admitted inside the vessel per second. The power of the signal arriving at the receiver input at the time $t = 0$ is equal to [see Eq. (62)]

$$P_{s,in} = \frac{\omega\mu_0^2\eta I^2 Q_L^2}{2\pi V\gamma^2 Q_c},$$
(63)

whence

$$I = \left(\frac{2\pi P_{s,in}\, V\gamma^2 Q_c}{\omega\mu_0^2\eta Q_L^2}\right)^{1/2}.$$
(64)

We determine the initial power of the signal by comparing it with the calibrated power of a noise generator, $2 \cdot 10^{-15}$ W. The signal power at the time $t = 0$ is equal to $1.5 \cdot 10^{-14}$ W = $1.5 \cdot 10^{-7}$ erg/sec. Substituting the numerical values of $\gamma = 3$ sec^{-1}, $V = 1.5 \cdot 10^4$ cm^3, $Q_L = 2.4 \cdot 10^4$, $Q_c = 6.4 \cdot 10^4$, $\omega = 8.9 \cdot 10^9$ sec^{-1}, $\eta = 3$, $P_{s,in} = 1.5 \cdot 10^{-7}$ erg/sec into Eq. (64), we find $I = 1.1 \cdot 10^{12}$ atoms/sec. The value obtained for the flux of atoms is near the value of the flux required in order for the oscillator self-excitation condition to be fulfilled.

2. Characteristics of the Hydrogen Quantum Oscillator

It is a well-known fact [30, 48] that transitions with $\Delta m_F = 0$ (σ-transitions) require the presence of a steady magnetic field whose strength vector coincides with the direction of the vector magnetic component of the high-frequency field $\mathbf{H}_\omega$ (for the mode TE_{011} at the center of the cavity $\mathbf{H}_\omega$ is directed along the axis of the cavity). If the constant magnetic field forms an angle θ with the cavity axis, the total fraction of atoms undergoing σ-transition is proportional to $\cos^2\theta$. The cavity axes of the apparatuses were not oriented in the direction of the external magnetic field, and the shields were not sufficiently effective, hence it was reasonable to expect that the oscillator self-excitation conditions at the transition $(F = 1, m_F = 0) \rightarrow (F = 0, m_F = 0)$ could be ensured by creating an additional magnetic field artificially in the direction of the cavity axis.

For creation of the axial magnetic field coils similar in construction to Helmholtz coils were wound on the resonator cylinders. Each coil had 35 turns of wire, and the coils were spaced about 14 cm apart in symmetric fashion with respect to the vessel. Beginning with a certain current in the coils, the rate of decay of the induced emission signal diminished, indicating narrowing of the line due to regeneration (Fig. 19). With a somewhat larger current the apparatus began to operate in the oscillation regime. The generated signal amplitude increased with the current in the magnetizing coils, reaching a maximum with a field of about 300 mOe in the vicinity of the vessel (for a current of 90 mA in the coils). For large values of the field the signal amplitude decreased due to the influence of inhomogeneities in the field produced by the

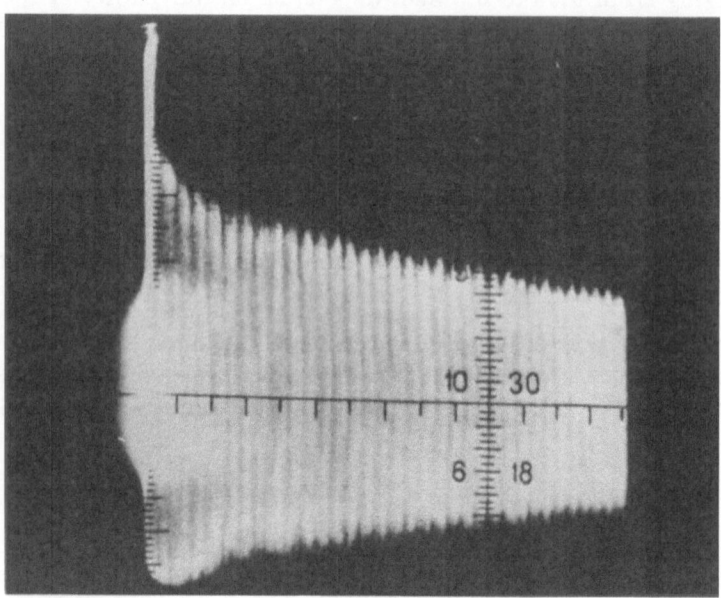

Fig. 19. Form of the induced emission signal on approaching
the self-sustained oscillation regime.

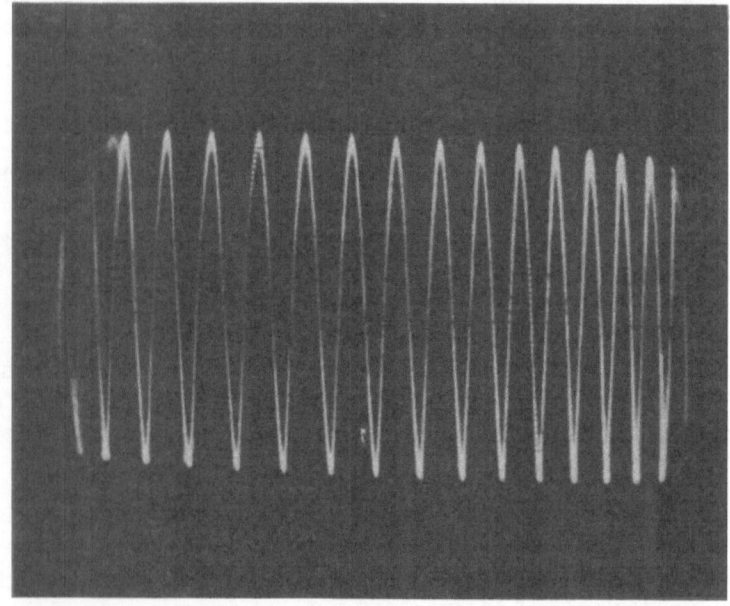

Fig. 20. Self-sustained oscillation signal. The signal power
is 10^{-14} W, the signal-to-noise ratio is greater than 20.

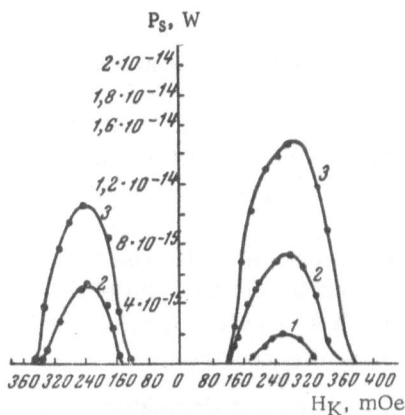

Fig. 21. Dependence of the oscillation signal power of apparatus No. 1 on the magnetic field H_K produced by the magnetization coils for various intensities I_V of a beam of atoms in the state $F = 1$, $m_F = 0$. 1) $I_V = 5 \cdot 10^{12}$; 2) $8 \cdot 10^{12}$; 3) $1 \cdot 10^{13}$ atoms/sec.

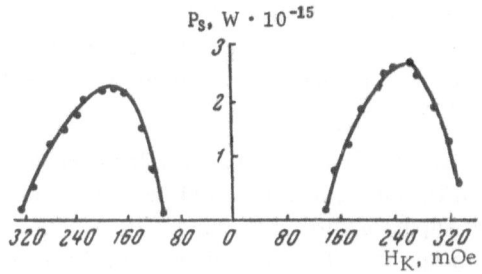

Fig. 22. Dependence of the oscillation signal power of apparatus No. 2 on the magnetic field H_K.

coils. The shape of the generated signal is shown in Fig. 20. The power of the signal admitted to the receiver was measured by comparison with the calibrated power of a noise generator, $2 \cdot 10^{-15}$ W. During operation without shields, this power was about $1 \cdot 10^{-15}$ W.

Three-layer cylindrical shields consisting of two layers of 79 NM Supermalloy 1 mm thick and an outside layer of Armco iron 3 mm thick were designed and fabricated for the elimination of any external magnetic fields. The shields were subjected to suitable heat treatment in a vacuum furnace. After installation of the first shield and the passage of optimum current in the coils, it was possible to increase the oscillation amplitude significantly (the signal at the receiver input increased to $2 \cdot 10^{-14}$ W). The second and third shields did not exert an appreciable influence on the self-excitation conditions or the amplitude of the emitted signal.

The dependence of the signal power at the receiver input on the magnetization field strength H_K for the first and second oscillators is shown in Figs. 21 and 22. As apparent from those figures, oscillation begins for different values of the field H_K, depending on the sense of the current through the coils. This is clearly attributable to the presence of the remanent field component H_{0z} in the cavity, directed along the z axis of the latter. We can estimate the value of H_{0z}. Clearly, self-sustained oscillation sets in for a definite value of the axial field:

$$H = H_{K+} + H_{0z} = H_{K-} - H_{0z}, \tag{65}$$

where H_{K+} and H_{K-} are the field strengths created by the coils with the current in them running in opposite senses. Hence, $H_{0z} = (H_{K-} - H_{K+})/2$. We find from Fig. 21 for oscillator No. 1 that $H_{K-} = 144$ mOe, $H_{K+} = 124$ mOe, so that $H_{0z} = 10$ mOe. By similar treatment of the curves shown in Fig. 22, we find for oscillator No. 2 that $H_{0z} = 20$ mOe.

Figure 23 shows the dependence of the self-sustained oscillation signal power on the beam intensity. The beam intensity was estimated from the value of the pressure in the first section of the vacuum system. Assuming the evacuation rate of the first section to be 200 liters/sec, and recognizing that the effective capture angle of the magnet $\Omega_{eff} = 1.3 \cdot 10^{-4}$ ster, it is possible to ascribe to each value of the pressure a definite value of the atomic flux into the vessel. The indicated calibration is made in the figure. It is evident from the graph that the oscillation signal power rises at first with increasing beam intensity, then drops rather quickly to zero. Oscillation cutoff is associated with line broadening due to atomic collisions (thereby causing spin exchange) in the vessel at sufficiently high pressure. At the instant of cutoff the pressure of the gas in the third section is equal to 8-$10 \cdot 10^{-7}$ mm Hg.

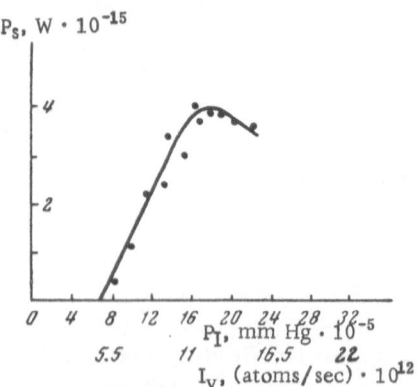

Fig. 23. Dependence of the oscillation signal power of apparatus No. 1 on the atomic beam. The horizontal axis represents the pressure in the first vacuum section and the approximate fluxes of atoms in the state $F = 1$, $m_F = 0$ admitted to the storage vessel.

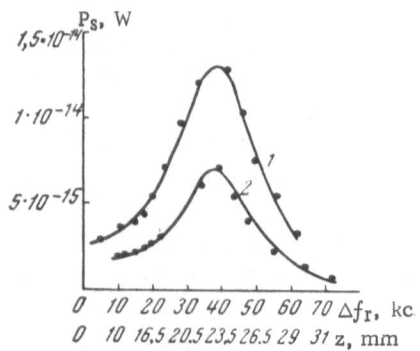

Fig. 24. Dependence of the oscillation signal power of apparatus No. 1 on the frequency separation of the resonator. The horizontal axis represents the separation Δf_r of the cavity in kc and the depth of immersion into the cavity of a tuning stick 8 mm in diameter (1). The amplitude of a 1420.4-Mc signal transmitted through the cavity for various frequency separations of the latter is shown for comparison (2).

The dependence of the oscillation signal power on the frequency separation of the high-frequency resonator relative to the spectral line is shown in Fig. 24. Also shown for comparison (curve 2) is the amplitude of a 1420.4-Mc signal transmitted through the cavity as a function of the frequency misalignment of the cavity. As the graphs indicate, the oscillation signal power drops to one half the peak value when the cavity is separated by one bandwidth (about 30 kc) from its resonance.

Figure 25 shows the dependence of the oscillation signal power of apparatus No. 1 on the current in the windings of the focusing magnet. In the interval of small current values (from 1 to 5 A) the signal power climbs rapidly due to the increasing effective capture angle of the magnet as the magnetic field is increased. With a further increase in current (ranging from 5 to 12 A) the signal power remains virtually invariant.

The frequency of the oscillators was measured as a function of various parameters by means of the arrangement shown schematically in Fig. 26.

The signals from the two oscillators were fed through ferrite gates to a hybrid loop and then to the input of the receiver. The 405-kc signal from the receiver output was delivered to a D2-E crystal diode, which discriminated the beat frequency of the two oscillators. After amplification, the signal was sent through a low-frequency filter to the input of the vertical-deflection amplifier of the ÉO-7 oscilloscope. The voltage from an NGPK-3 low-frequency oscillator was applied simultaneously to the horizontal sweep plates. The beat frequency of the two oscillators was measured by observation of the Lissajous figures created on the oscilloscope screen. The oscillator frequency did not vary more than 0.1 cps for a 20% variation in the beam intensity due to variation of the pressure in the source. The frequency also fluctuated slightly with variations of the current in the windings of the focusing magnet and with a reduction in beam intensity due to shielding.

With an approximately 30-kc readjustment of the resonator frequency by means of a tuning stick, the oscillator frequency varied by about 0.5 kc; this means that the width of the resonance line was about 1 cps. The oscillator frequency depended most strongly on the external magnetic field. The curves in Figs. 27 and 28 show the dependence of the oscillation frequency on the

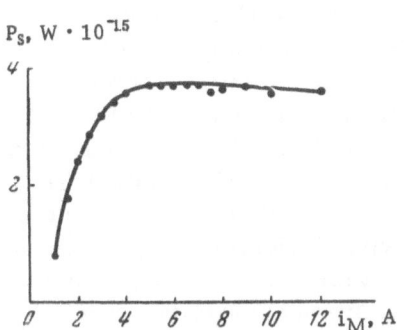

Fig. 25. Dependence of the oscillation signal power of apparatus No. 1 on the current i_M in the windings of the six-pole focusing electromagnet.

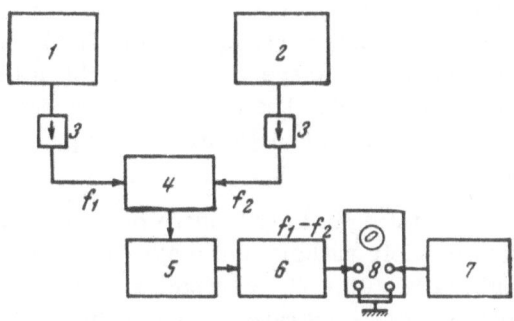

Fig. 26. Scheme for comparison of the frequencies of two quantum oscillators. 1) Oscillator No. 1; 2) oscillator No. 2; 3) ferrite decouplers (20 dB); 4) hybrid loop; 5) receiver; 6) detector and low-frequency amplifier; 7) NGPK-3 low-frequency oscillator; 8) ÉO-7 oscilloscope.

applied field. In recording the curves, the frequency of one of the oscillators was set at a fixed value, while the other was varied by changing the magnetic field in the Helmholtz coils. The frequency differences of the two oscillators were plotted on the ordinate axes.

The curves are similar to parabolas, but not quite. The right and left branches of the curve are asymmetric with respect to the axial line. Another technique for determining the projection H_{0z} of the remanent field on the cavity axis is to measure the degree of incongruity of the curves. As a matter of fact, the oscillator frequency when a remanent magnetic field H_0 with components H_{0z} and $H_{0\perp}$ present in the cavity must depend as follows on the additional magnetic field applied parallel to the axis:

$$\nu_g = \nu_0 + 2750 \,[H_{0\perp}^2 + (H_{0z} \pm H_{\kappa})^2] = \nu_0 + 2750 \,[H_0^2 + H_{\kappa}^2 \pm 2H_{\kappa}H_{0z}], \qquad (66)$$

where ν_0 is the transition frequency in the absence of any external magnetic field, the plus and minus signs in front of the last term refer to the right and left branches of the curve, respectively. For a fixed value of $|H_K|$ the difference between the functions ν_g for the two branches is equal to

$$\delta\,(\nu_g) = 11.000 \, H_{\kappa}H_{0z}. \qquad (67)$$

Plotting the straight line (67) and finding its slope, we thereby determine H_{0z}. For oscillator No. 1 we have $H_{0z} \approx 10$ mOe; for No. 2 about 20 mOe. This is consistent with the previous estimate.

For calibration of the horizontal axis directly in oersteds, we constructed the parabola described by the law $\Delta\nu = 2750 \, H_K^2$ and shifted relative to the experimental curves by an amount $\pm 2H_K H_{0z}$. Segments of these parabolas are indicated by dashed curves in the figures.

It is not possible to process the curves in order to obtain the value of the remanent field; all that can be done is to ascertain the difference between the squares of the remanent field values for the two oscillators, $H_{01}^2 - H_{02}^2$. For determining the remanent field it is best to use the method of induced transitions with $\Delta F = 0$, $\Delta m_F = \pm 1$ [12, 22]. If the cavity is made of a dielectric (quartz, for example), a resonance field is easily established by placing the coils outside the cavity [22]. This method is not applicable for a metal cavity. An audio frequency field is set up by means of a coupling loop placed inside the cavity. The loop, which was first proposed and tested by G. and L. Elkin, consists of two segments of silvered wire situated diametrically opposite one another along generatrices of the resonator cylinder and at a distance of 1 or 2 mm from its inside surface. Both halves of the loop are connected by a wire running

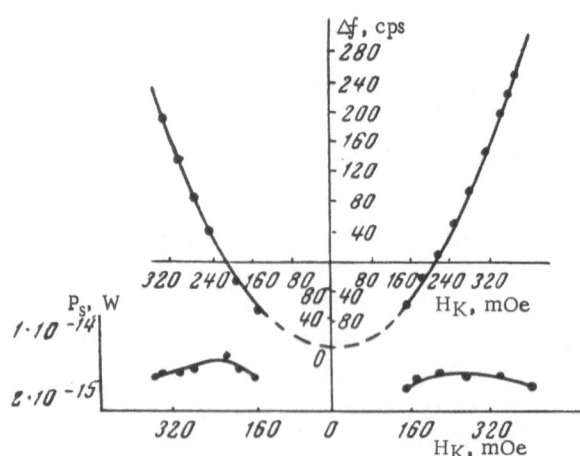

Fig. 27. Dependence of the oscillation frequency of apparatus No. 1 on the magnetic field H_K produced by the magnetization coils. The amplitude characteristics are shown in the lower half (see Fig. 25).

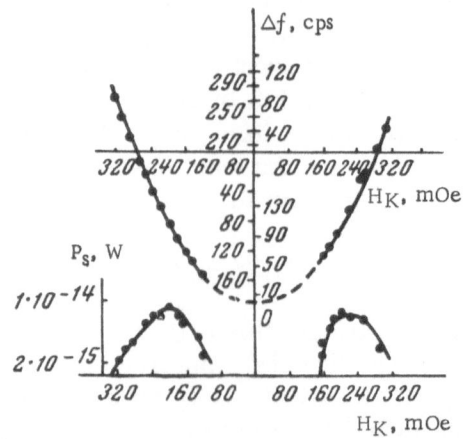

Fig. 28. Dependence of the oscillation frequency of apparatus No. 2 on the magnetic field H_K.

outside the cavity; this type of loop does not particularly diminish the Q of the resonator. In apparatus No. 1 with a current of 90 mA in the Helmholtz coils (at which the oscillation signal has peak amplitude) a reduction in signal power begins when a signal with a frequency of 402.5 ± 0.4 kc is supplied to the inside of the cavity, indicating the presence of a 288-mOe field therein. At a frequency of 400 kc, the π-transition absorption linewidth is equal to 30 kc. Such a broad line ($\Delta m_F = \pm 1$) implies a high degree of inhomogeneity on the part of the constant magnetic field in the cavity.

By measuring the transition frequency $|\Delta m_F| = 1$ for various values of the current through the Helmholtz coils and plotting a graph of this dependence (a straight line), one is able to determine the remanent field in the absence of current. The remanent current turns out to be equal to 10 mOe for apparatus No. 1.

The distinguishing feature of both apparatuses is their rather high magnetic field strength (100-160 mOe) at which self-sustained oscillations are generated. The attainment of an oscillator stability of $1 \cdot 10^{-12}$ requires that the relative instability of the current in the power supplies to the magnetizing coils be equal to

$$\frac{\Delta H_{\text{K}}}{H_{\text{K}}} = \frac{10^{-12}\nu_0}{5500 H_{\text{K}}^2}. \qquad (68)$$

For $H_K = 1.5 \cdot 10^{-1}$ Oe, the ratio $\Delta H_K / H_K$ must be at least $10^{-3}\%$, but for $H_K = 5$ Oe it is only necessary in order to achieve a stability of $1 \cdot 10^{-12}$ to hold the field in the Helmholtz coils constant within a 1% margin. The limit below which it is no longer meaningful to reduce the magnetic field is determined by the requirement that the levels F = 1, $m_F = 1$, and F = 1, $m_F = -1$ fall outside the resonance linewidth (1 cps). In order to shift the frequencies of these components, say, 30 cps relative to the line (F = 1, $m_F = 0$) → (F = 0, $m_F = 0$), it is necessary in the region occupied by the vessel to have a constant field $H_K = 2 \cdot 10^{-5}$ Oe. The factors dictating the need for operating with a relatively high field H_K are the following: a) The quality of the shielding is not quite adequate; b) the Helmholtz coils (for engineering reasons) are too close to one another and generate a uniform field only in a portion of the vessel interior; c) the brass used in the construction of the cavity has a specific magnetic susceptibility of about 10^{-5}, which is about 100 times the specific susceptibility of copper. The field generated by the brass resonator in the vicinity of the vessel may be as high as several millioersteds and has a significant inhomogeneity.

After demagnetization of the inner shield (by passing a gradually diminishing 50-cps current through a special winding), the frequency of the oscillator operating with a magnetizing field $H_K = 260$ mOe varied by 4 cps. This corresponds to a field variation of 3 mOe.

Generally speaking, there are several reasons for the frequency shift in the atomic hydrogen quantum oscillator.

1. Frequency shift due to the interaction of atoms with the wall (the experimental value for the relative shift is $2 \cdot 10^{-11}$ [24], although it is not clear how this is affected by aging of the inner coating).

2. Frequency shift due to the first-order Doppler effect (the shift is negligible.

3. Frequency shift due to the second-order Doppler effect. This shift depends on the temperature and may be substantial. If the wall temperature is held constant within $\pm 0.15°C$, the relative shift will change less than $1 \cdot 10^{-13}$.

4. The dependence of the oscillation frequency on the resonator frequency and the transition frequency is given by the well-known relation

$$\frac{\omega - \omega_0}{\omega_0} = \frac{\omega_r - \omega_0}{\omega_0} \frac{Q_r}{Q_l} .$$ (69)

If we do not want the relative shift to exceed $1 \cdot 10^{-13}$, we must have

$$\frac{\omega_r - \omega_0}{\omega_0} < 5 \cdot 10^{-9}.$$

5. The magnetic field is not quite constant. For a relative frequency stability no worse than $1 \cdot 10^{-13}$ it is necessary to hold the 10-mOe field constant correct to $2.5 \cdot 10^{-2}\%$.

6. The presence of molecular hydrogen in the vessel [16], along with other atomic hydrogen states (F = 1, $m_F = \pm 1$).

The principal factors limiting the stability of the quantum oscillators constructed for these experiments are instability of the constant magnetic field and a deviation of the resonator frequency due to thermal expansion, as well as fluctuations in the pressure, temperature, and moisture content of the air in the cavity. The first factor must be removed by striving to diminish the level of the remanent field through improved shielding quality, and to use systems creating a highly stable and homogeneous constant magnetic field. For elimination of the second factor it is essential to construct the resonators of a material with a low expansion coefficient, and then to provide them with good thermostatic regulation and evacuation.

CONCLUSIONS

In the present study we have described the structure and tuning of a quantum oscillator device operating on a beam of hydrogen atoms and utilizing the transition (F = 1, $m_F = 0$) → (F = 0, $m_F = 0$) at a line frequency of 1420.405 Mc.

Two apparatus modifications differing in their construction were fabricated and tested. The behavior of the quantum oscillator in the subexcitation regime was investigated. A procedure was described for determining the lifetime of the excited atoms in the storage vessel. The self-sustained oscillation regime was initiated with a constant axial magnetic field of 100-300 mOe in the resonator. The oscillation amplitude and frequency were investigated as a function of various parameters. It was found that instability of the additional constant magnetic field and misalignment of the cavity resonator due to thermal expansion provide the major contribution to the over-all instability. Possible ways in which these principal causes of instability in the oscillation frequency might be removed were suggested.

The authors express their appreciation to A. M. Prokhorov and A. N. Oraevskii for a discussion of the results and valuable advice, as well as to L. P. Elkina, G. A. Elkin, A. N. Ponomarev, A. A. Ul'yanov, L. M. Zak, N. A. Begun, and O. S. Lysogorov for assisting with the study.

LITERATURE CITED

1. N. G. Basov and A. M Prokhorov, Zh. Eksperim. i Teor. Fiz., 27:431 (1954).
2. J. Gordon, H. Zeiger, and C. Townes, Phys. Rev., 95:282 (1954).
3. J. Gordon, H. Zeiger, and C. Townes, Phys. Rev., 99:1264 (1955).
4. N. F. Ramsey, Phys. Rev., 78:695 (1950).
5. J. R. Zacharias, Phys. Rev., 94:751 (1954).
6. N. F. Ramsey, Rev. Sci. Instr., 28:751 (1954).
7. R. H. Dicke, Phys. Rev., 89:472 (1953).
8. H. Goldenberg, D. Kleppner, and N. Ramsey, Phys. Rev., 123:530 (1961).
9. D. Kleppner, N. Ramsey, and R. Fjelstadt, Phys. Rev. Letters, 1:232 (1958).
10. J. R. Wittke and R. H. Dicke, Phys. Rev., 103:620 (1956).
11. H. M. Goldenberg, D. Kleppner, and N. F. Ramsey, Phys. Rev. Letters, 5:361 (1960).
12. H. M. Goldenberg, Thesis. Harvard University, Cambridge, Mass. (1960).
13. W. Gordy, W. V. Smith, and R. F. Trambarulo, Microwave Spectroscopy. Wiley, New York (1953).
14. H. Jonson and M. Strandberg, Phys. Rev., 85:503 (1952).
15. C. Menoud and J. Racine, Helv. Phys. Acta, 35(7-8):562 (1962).
16. D. Kleppner, H. Goldenberg, and N. Ramsey, Phys. Rev., 126:603 (1962).
17. N. G. Basov, Doctoral Dissertation, FIAN (1956).
18. A. N. Oraevskii, Molecular Oscillators. Izd. Nauka, Moscow (1964).
19. K. Shimoda, T. C. Wang, and C. H. Townes, Phys. Rev., 102:1308 (1956).
20. N. Ramsey, Microwave J., 6:89 (1963); IRE Trans. Instr., 1-2:117 (1962).
21. R. Vessot and H. Peters, IRE Trans. Instr., 1-2:183 (1962).
22. D. Kleppner, H. Goldenberg, and N. Ramsey, Appl. Opt., 1:55 (1962).
23. H. Berg and D. Kleppner, Rev. Sci. Instr., 33:248 (1962).
24. S. Crampton, D. Kleppner, and N. Ramsey, Phys. Rev. Letters, 11:338 (1963).
25. P. Grivet and N. Bloembergen, Quantum Electronics, pp. 333, 409. Proc. Third Internat. Congress, Paris, 1963 (Paris, New York, 1964).
26. J. Vanier and R. Vessot, Appl. Phys. Letters, 4:122 (1964).
27. A. S. Nazarov, G. R. Ivanovskii, and M. I. Men'shikov, Pribory i Tekhn. Eksperim., 5:157 (1959).
28. B. S. Danilin, Design of Vacuum Systems. Gosenérgoizdat (1959).
29. L. Paty, Pribory i Tekhn. Eksperim., 6:1 (1959); A. M. Grigor'ev, Pribory i Tekhn. Eksperim., 6:10 (1959).
30. N. F. Ramsey, Molecular Beams. Clarendon Press, Oxford (1956).
31. Polarization of Nucleons [Russian translation]. IL, Moscow (1963).
32. V. B. Leonas, Usp. Fiz. Nauk, 82:287 (1964).
33. B. P. Ad'yasevich and V. G. Antonenko, Pribory i Tekhn. Eksperim., 2:157 (1963).
34. J. Estermann, Rev. Mod. Phys., 18:300 (1946).
35. W. A. Lamb and R. K. Riserford, Usp. Fiz. Nauk, 45:553 (1951).
36. A. M. Bass and H. P. Broida (eds.), Stabilization of Free Radicals at Low Temperatures. U. S. Dept. of Commerce, Nat. Bur. Standards, Washington, D. C. (1960).
37. K. Landecker and A. Gray, Rev. Sci. Instr., 25:1151 (1954).
38. E. Harrison and L. Hoblis, Rev. Sci. Instr., 26:305 (1955).

39. E. T. Kucherenko and O. K. Nazarenko, Pribory i Tekhn. Eksperim., 6:124 (1959).

40. R. W. Wood, Proc. Roy. Soc., 97:455 (1920).

41. V. P. Strunin and E. L. Frankevich (in press).

42. M. I. Korsunskii and Ya. M. Fogel', Zh. Eksperim. i Teor. Fiz., 21:25 (1951).

43. H. Friedburg and W. Paul, Naturwissenschaften, 38:159 (1951).

44. H. Friedburg, Z. Phys., 130:493 (1951).

45. H. Beenwitz and W. Paul, Z. Phys., 139:489 (1954).

46. A. Lemonick, F. M. Pipkin, and D. R. Hamilton, Rev. Sci. Instr., 26:1112 (1955).

47. R. L. Christensen and D. R. Hamilton, Rev. Sci. Instr., 30:356 (1959).

48. K. F. Smith, Molecular Beams. Methuen, London; Wiley, New York (1955).

49. D. D. Chegodaev and N. E. Yavzina, Application of a Teflon Suspension, Izd. LDNTP (1960).

50. D. D. Chegodaev, Z. K. Naumova, and I. S. Dunaevskaya, The "Ftoroplasts," (polyfluoro-ethylene resins). Goskhimizdat, Moscow (1960).

51. A. A. Lapis, Cavity Resonators. Izd. LGU, Leningrad (1954).

52. Ya. D. Shirman, Microwave Guides and Cavity Resonators, Svyaz'izdat, Moscow (1959).

53. V. B. Shteinshleiger, Wave Interaction Effects in Electromagnetic Resonators. Oborongiz, Moscow (1955).

54. A. G. Gurevich, Cavity Resonators and Waveguides. Izd."Sovetskoe radio," Moscow (1952).

55. A. P. Sivers, Radar Receivers. Izd."Sovetskoe radio," Moscow (1959).